# BRIDGES

*Getting from A to B*

# BRIDGES

## Getting from A to B

MARK GEPPERT

GENESIS

Published by Genesis Books
An imprint of ARMOUR Publishing Pte Ltd
Kent Ridge Post Office
P. O. Box 1193
Singapore 911107
Email: armour@singnet.com.sg

Cover Design by Flora Chan
Cover Photo by Davis Yap



Printed in Singapore

ISBN 981-4045-30-6

# Contents

# Dedication

EACH one of us is the product of the inputs of others. I have especially respected and treasured the inputs referred to in this book. While some of the stories are based on actual people and places, certain details have been changed or omitted entirely so that no one person is singled out.

The issue at hand is the system by which so many of us have been stumbled in our attempt to serve the Lord. As you read this book, I trust that you will have compassion on those mentioned.

I would like to thank those who traveled in the van from Thailand through Vietnam reading the early manuscript. As you can see, I adopted all of your suggestions. I especially would like to thank my sister Christine for editing the final copy. Now I understand why one of the children in our family had to be a girl who studied and got good grades in English. All things are made clear over time.

I mention the Rev. Dr. Naomi Dowdy quite a bit in this book. She has become an inspiration to me as I feel that we are at the dawning of the day when women will be recognized in leadership in the church. Her style adorns the Gospel, which she preaches.

Bob and Lesley continue to be my inspiration. They are in every church. They serve and are hurt and serve again. They do not allow themselves to be overcome by emotion. They love God even when His children misbehave. This one's for you.

Then there is Ellie. She has been a Pastor's wife. She has been a Pastor's kids' mother. She is herself a wonderful minister of the gospel. In case you wondered, she is not Lisa Greene. Thank you for listening so well all these three decades. May the Lord bless you.

And then there is you, the reader. It is my fervent prayer that this book helps you to get from A to B in God. These principles have really helped me. I have learned them by living through some tough times. I hope you can learn and apply them. If so, I have succeeded.

*Chapter 1*

# Disaster

RED, blue, red, blue ... staccato strobes of white light danced their silent echoes down the narrow side street. They seemed to be announcing his sin over and over, causing him to slide deeper into the dark recesses of the patrol car. The handless door and bulletproof tinted window became barriers against the stinging accusations of the outside world of which until a few minutes ago he had been a part of.

Nothing could shield his mind from the acid realization that life, as he had known it, had just ended. His lungs ached with the agony of that realization, and his palms were damp with fear as he felt the plastic restraint cutting into his wrists. It would have been foolish to try to resist. He had known that one day he would be caught. In fact, the thrill of the secrecy of these midnight liaisons had been a big part of the bondage which he had been drawn into.

There had been warnings. Even the local police had told him just months before that, as a Pastor, they would allow his offenses to slide by, but he would have to find help and not be found on these streets again.

Perhaps that was the greatest thrill. In his world of local church, he was the leader. No one there could seriously challenge his power and authority. He was the one, after all, who had signed for the construction loan to build their new million-dollar facility. The loan had been based on his ability to fill the place. People came from as far as 50 miles to hear him teach the Word of God on Sundays. They drove past many other churches to get there.

The church was known as "Pastor Robert Greene's" church. It bore his name, not the names of those laymen who served as a board. When parishioners spoke of the church, they attached his name. He was very highly thought of in the community where he served on many committees for community growth.

In the 20 years he had been there, the church had grown from a home group of ten into a significant congregation. Along the way, there had been those who cautioned that he had not made himself accountable to anyone. As these people left, they had gently suggested that he join a larger ordaining body with real accountability, but he had seen them as malcontents. Some of them had even suggested that there might be a problem in his marriage.

His marriage; how was Lisa going to handle this? Solicitation of sexual favor from an undercover policewoman — would it be the wedge that would finally drive them away from each other? Not that they had been close recently. The fire of their marriage had cooled to an ember after the third child. Megan had brought with her a package of problems.

His children would have to face the morning news at school. What would he be called by their taunting peers? How could they face the pressure? How many times had he lectured them on sexual purity? They had come forward and taken the pledge of purity during the last youth rally. Jainine had wept as she thought of saving herself for the Lord's best; James had testified of the release of a great pressure in his heart as he prayed the prayer. Even at 12 he was feeling the pressures.

How would they handle this — that their father was a whoremonger, a fraud, and a felon. He must reach them before morning to tell them it was all a mistake, a once-in-a-lifetime fall from God's grace. He had to get there before the reality of his sin crushed their idealistic world of lace and model cars. Why was this taking so very long? Couldn't it be sped up? This moment would, like all others, pass with a pleasant apology and pastoral cool.

"Sir," the officer interrupted his musing. "We are taking you to the night court. You are under arrest for solicitation of sex. This is a felony charge in this State. You have the right to remain…"

The chill around his heart became ice. His breath came in gasps. This was really happening. He was going to be exposed, ruined, disrobed. His family would be destroyed and his ministry finished. There was no defense he could offer. He could not charm his way out. He had no authority in this back seat and, as the car left the narrow street and sped along the major highway, life was left in the darkness.

"Reverend, how is it that I find you in my court tonight?" The Justice had been prepared for this preplanned sweep of the "red light district." The police had told him that they were watching several known prostitutes and had managed to get a vice officer into their midst. They had warned several of the more prominent citizens to stop their wandering ways, but this was the time that they had to clamp down on the area. Those caught in the net would have to be made an example.

The Justice had thought of calling the Pastor to let him know that they were aware of the frequency with which he visited the street and would "close one eye" for him if he did not return. The Justice was very aware of the good which had been done by the church and the possible impact on it and the community if the police followed through with their arrests. The police chief and he had discussed the sensitive nature of this before agreeing that a standard had to be set. They had determined that if a warning went unheeded they would have to bring to bear the full impact of the law.

Two weeks before they had been made aware that the Pastor was on the street. They had an officer approach him and instruct him to leave the area. He had tried to say that his being there was coincidental. The officer had reported to the Chief, the Chief to the Justice, and the decision had been reached.

For the first time in decades the Pastor was speechless. Any answer he could give would seem very empty. Remorse was strange for him. No one challenged his authority at work or at home. He was able to talk his way out of or into any situation. He had no tools left this moment.

"I suggest you call an attorney," the Justice followed normal procedure. "You may use the public phone. If you do not have an attorney, one will be provided for you."

Alone, this is what it feels like to be alone. Who could he call? Who could he bring into this night of personal horror? Which of his trusted friends could walk with him through this? Bewildered, he looked to the Justice. So many times he had worked out arrangements with this man to help another in his position. They had managed the restoration of perhaps a hundred people from the community. Now he was in the position of need. This was not a back-room meeting or a phone call on behalf of another. This was his life.

"Isn't there some way we can work this out?" he asked.

"You have been charged before this court with felony solicitation. This is a charge which, if you are found guilty, can result in a very stiff fine and jail time. It can result in a felony conviction being entered into your records, the consequences of which are very serious. You need legal counsel before we continue. I cannot make a decision on the expediting of your case until you have legal counsel. You must either call your attorney or we can appoint a public defender, but you must make the decision." The Justice did not smile or give any sign of the years of *relationship. This* was business.

Walking to the offered phone he saw the reporters. These were not the ones with whom he had worked each year to announce the Christmas and Easter presentations. Nor was it his old friend who did the annual presentation of his testimony and the plug for the church. These were out for blood. They even took his picture and rushed out with the charges in hand. It would be on the TV news before breakfast. Bread, bagel and Robert Greene would all be toasted before seven. He was finished.

Release came at nine. The Justice had let him go home under his own recognizance. His attorney had gotten out of bed and come immediately and was walking with him to his home. They had decided that he would call home himself and just say that there had been a problem and that he would be home as soon as possible. He did suggest that the kids stay home from school, as it was quite serious.

Lisa met them at the door. Her tear-streaked face told it all. Defying his instructions, she had seen the TV news and was well aware of the charges and certain guilt. As he came through the kitchen door, she kept the counter between them. The kids had also heard. They were in shock.

The church elders would be there at 11. They were meeting and would let him know at that time where he stood with them. Friends had called to encourage her and family had called to see if it was as reported. Their support meant a lot to her. The kids could not believe it.

Several TV stations had called and asked for interviews. She had declined saying that the church would offer a statement around noon. Two Pastors had called and offered their help in prayer and counsel. They were well respected and ready to come to their side. Just give them a call if need be. Each had left his personal handphone number. Megan was having difficulties again. Her breathing was not clear. She would need to be taken to a doctor today.

Lisa suggested he shower and shave and get ready for a long day of questions and answers. He nodded and carried the weight of the world up the stairs to their room. The kids heard his footfall, but did not come to greet him.

Water could not wash the stain from his heart. He scrubbed as hard as he could, but the washcloth could only touch the skin. His guilt lay much deeper. Finally, tears mixed with steam and the cleansing process began. All he would do all day was ask forgiveness. He quietly prayed and placed his future in the hands of the One who had not let him go. It was Jesus who had allowed the law to bring him to grace and it would be Jesus who would lead him on. He called upon Jesus to touch him and felt the presence of the Lord with him. Arrogance washed down the drain with his reputation and tears cleansed his soul. He stood naked before the Lord, a man stripped of title and position, livelihood and power, and ashamed and broken he wept as never before.

He called the Pastors first. Yes, it was true. Yes, he asked for their help. No, he was not at all prepared for the trial that stood before him. Yes, the elders of the church were meeting and would be over. No, he and his wife had not had the chance to sit alone and talk about it. No, he had not seen his kids. No, they had not gone to school; they were in the house somewhere. Yes, please do come over.

Jainine and James came through the door together. They gave him big hugs and James said it couldn't be true. He told them it was. He had been unfaithful to them and to their Mom and he asked them to forgive him. They did not. They looked at him in shocked disbelief. Ready to defend, they stood now completely disarmed. Their hero sat in a pool of filth so deep that they could not hear his cry for help. They turned and ran to their Mother in tears.

As he followed he could hear Megan gasping in her room. The breathing was really labored; they would soon have to go out to the doctor. Maybe just this once, given the circumstances, he would come. To make the call, he had to work through the secretary who had also seen the morning news. His call was rushed through to the doctor. The doctor was in shock and assured them of his friendship and availability. He would come right over to care for Megan. He arrived at the same time as the two chariots of conviction entered the drive.

This was worse than the Justice. He was in the business of looking at the guilty. It was his profession. They were tradesmen in the community whose greatest issues had been the color of the carpet in the new sanctuary. They had never had to deal with such a thing. Their Pastor was caught in the act in the middle of the town he had just taught them to reach with the purity of the Gospel. They were hurt. They were angry. They were disappointed. Healing went to Megan's room and Judgement convened in the family room.

They had appointed Luke, a salesman, to be the spokesman for them all. They prayed together and then he delivered their joint statement. They felt that since the charges were confirmed, and the community made aware, that they should issue a statement through the media that they loved their Pastor and would stand with him and his family through this unfortunate situation. They recommended that he submit himself to a small group of local clergy for the purpose of restoration. They assured him that his salary and benefits would continue for one year and that after that they would reassess. They felt it was a serious blow to the church which would take several years to heal, but they remembered all he had done for them and were going to stand with him in his time of need.

He was overwhelmed by their love. He wept before them asking their forgiveness for his sin. He thanked them for coming. Luke would call the papers and TV from there and give them the statement if Pastor was agreeable to the small group accountability for restoration. He embraced the idea and told them of the calls to the two Pastors and their plans to come to the house that afternoon. Please make the call so that it would be on the noon news shows and the whole process could move along.

Lisa watched from the doorway. Her emotions were caught up in a very large knot high in her throat. She did not know if she would vomit away the extreme distaste she had for him or if she would choke to death from embarrassment. While they were attending to Megan the Doctor had asked her if she needed anything for her nerves. He would be glad to prescribe something. She had assured him that she was fine and had even said that the grace of the Lord was keeping her. Now that grace seemed very far away. Should she move to his side? Nothing in her wanted to. Should she offer coffee and tea to the elders? Her usual wifely servant role in their meetings felt somehow stained as she watched her husband's shoulders sag more and more.

His strength had been the force that had drawn her to him. In the adventure years they had borne up under tremendous pressure.

Money had been very tight, but they had found joy in sharing soup and a hamburger. He had even refrained from ketchup when they ordered fries because he knew she didn't like it. When the kids came, the shoulders were their favorite riding place. Their strength had carried Jainine and James high above the pressures of life. They had learned the love of God through the strength of their Father. Often he had said to them that God would lift them up to His shoulders, imagine how high that would be!

Megan's fibrosis had put a little sag in them. It was beyond his ability to change the situation, but he had rebounded and they had worked it out together and their faith had carried them. Now, as she looked from safety, the shoulders slumped. He was broken.

She walked across the room and put a hand on his right shoulder. She touched him. He wept as never before. She would stand by him. She would not leave. She would rally the kids. They had faced 20 years together and one flaw was not going to destroy all they had built. She would face it with him.

He did not turn to face her. Maybe he would later. Maybe when they were with the Pastors. Maybe with the kids. Alone, he could not yet do it. But a time would come. There would be a resurrection. Her touch confirmed to him that his life would not end on this most miserable of days. There would be very difficult issues to face, but together they would make it. She had touched him.

The elders in this all too real story had struck upon a very important truth, which the Lord has restored to the church in these days. They realized that the way to strength for today's challenges is a small group, closely knit together in relationship. Their Pastor had not had this kind of relational group and was set up for disaster. Unfortunately, his experience has been repeated in fearful numbers around the world. In this book we are going to address seven principles which will restore those who are in such crises. We will follow Robert Greene and his family through their situation. We will also look in on other ministries, which have built their bridges to the future.

We have called the book BRIDGES, How to Get from A to B. The letters of the title form the acrostic of the seven main points, Bases, Resources, Intercession, Discipleship, Going, Evaluation, and Sustaining Ministry. We are going to draw heavily upon the experience of the Rev. Dr. Naomi Dowdy and a wealth of other successful ministries. We trust that this book will help you to restore those in your world who have fallen. More especially, we hope it will prevent a fall.

*Chapter 2*

# The Way Back

BUILDING a relational base for your ministry is vitally important to your success. There is wisdom in the multitude of counsel. There is safety in the multitude of counsel. War is waged in the multitude of counsel. To limit your vision, purpose, goals, and evaluation to your own point of view leads to the type of situation Robert Greene faces. Each of us needs counselors. The question is where do we find them?

## The Pyramid Structure

David sought the counsel of Samuel. Absalom sought the counsel of his peers and from them received the usual justification for his actions. Those who had gone before Him through doorways to eternity counseled Jesus. Robert Greene lived at the top of a pyramid each stone of which was his own opinion.

The classic view of leadership places a strong man at the top, a few lieutenants to do the bidding of the leader and a pyramid of those who, without question follow after. Of course, it takes a pillar of fire to keep them there. Yes, and a couple million Egyptians to get them to move out in the first place.

Moses led them out. He was a reluctant leader, given to outbursts of anger, capable of murder. He had spent a wilderness exile to be made a leader and now was walking in an impossible assignment. He needed help. He resorted to God for that help, but he knew he needed assurance and counsel from a friend. He suggested his brother, but that never seemed to work well. Finally it was his father-in-law who came to his side with a seemingly good suggestion, "Be a judge in large matters and create other judges for the small ones."

This sounded reasonable to a man who was frustrated with those he led. So Moses implemented the suggestion. He never stopped to ask the big question.

Who was Jethro anyway?

For many years this question wandered around the recesses of my mind. Each time I heard organizational teachings about the forming of pyramid structures, the question would be there. In the early '70s, while on a men's retreat, we listened to tapes by a young pastor from a successful church on the US East Coast. We followed along as he taught from Exodus 18 about the wisdom of Jethro as he counseled Moses concerning the distribution of the pressures of ministry.

We were very excited to hear of the appointment of an Eldership that would decide the small issues so that only the greater issues were brought to Moses. This of course gave Moses more time at home, and so his father-in-law, Jethro, did not have to hear his daughter's complaints that her husband spent too much time at the church and not enough at home.

We bought it. We agreed that from our midst elders should be appointed and that in this way the pressures of the ministry could pass from one pastor or "shepherd" in this case to many "under-shepherds." We sang songs, went home, grew like a weed and, for about six years, had a great time together.

Fissures began to appear. The relationship between those who were appointed and the pastoral team was strained at best. And by the end of the seventh year the entire structure had come down. In the midst of the acrimony and verbiage one thing was clear. **The pyramid scheme of Jethro did not work for us.**

About that time, I checked back on the young minister whose tapes we had digested. He had left the ministry after an affair with his secretary and was in the insurance business. Evidently, the structure had not worked for him either.

## What Is Wrong with the Pyramid?

Examination of the many scriptures concerning Midian and the Midianites revealed them to be antagonistic toward Israel. They were the sons of a concubine sent away by Abraham with a cash settlement (Genesis 25). Moses joined with them through marriage (Exodus 2:15). But the union was anything but bliss. Contention, which is the root of the name "Midian", seemed to follow all the days of the relationship between Midian and Israel. In Numbers and Judges, there is enmity between them.

Jethro was of the line of Abraham, but not the line of promise. His roots were in the concubine. His counsel was born of the conflicts which had come to pass among the followers of Moses.

As I observed these things and the outcomes of ministries which were locked into a pyramid structure, I began to see a pattern. Moses, at the top, became a very isolated individual. He was not able to lead the people to a promise of their own, but used them to obtain his own ends. Even though he coveted the presence of God, and received His word directly, his anointing diminished before the people. Why, I often wondered, did the man who received the tablets on the mountain, receive his administrative model from the Midianites?

After observing ministries these past 25 years, I have come to a conclusion. Pyramids are the sign of Egypt. They are the death we are to leave. They are the desert wasteland to which Israel was sent to cleanse her for her journey to the Promised Land. They are the burial grounds of would-be kings whose reigns are at the expense of the people. They will put the Lord's people into bondage and hold them there for generations.

But, why did God remain silent in this case and allow the counsel of Jethro to develop a Midianite structure for Israel? It made no sense to me. Certainly delegation is a part of successful leadership. Why didn't God give a design to Moses? What was the problem with this model? For many years I struggled as I watched men of God change from loving, compassionate, caring individuals to minimoguls in little empires competing viciously with each other.

From Louisiana to Kansas City they fell with loud crashes. They cheated on wives, stole money, lied to each other and themselves. Even when the federal government stepped in and made public their fall, they finger pointed and placed blame. The church was coming to a shambles as compromise rushed in to fill the void of departing leadership.

Those who collapsed had a common understanding. Every corporate structure was drawn in the shape of a pyramid. They were Moses with two trusted lieutenants faithfully executing their wishes and demands. A survey of ministries in the '80s shows the pyramid structure throughout. And it shows a trail of flashing lights, IRS seizures, public disgrace and denial.

The Pyramid does exactly what it was designed to do. It kills. That night in a small town, Pastor Greene felt the chill of its killing power. His children felt the pangs of the death of their hero. His wife refused to die, and thank God for his elders who refused to sound the death knell for the man they loved and respected. Moses was not so fortunate.

In a pyramid structure, authority is delegated down-line. It usually flows from the Senior Minister through two or more subordinates to the ministry heads. These subordinates have a very key position for they report to the head the things they want him to know in the way they want him to know them. Did Jim Bakker know how the money was spent? Did Jimmy Swaggert know all the details of his empire? They probably did not.

The "second fiddles" realize that to be first they must bide their time and control information. Information is power. They become power brokers. Theirs is a world filled with bits and pieces of information which they pass on only at key times. Aaron was so faithful until the people needed an idol. Then his true colors were manifest. Given the opportunity to rise in the eyes of the people, he made an idol and they danced to his tune. That dance was not new that day; his position among them had been developed long before that fateful manifestation.

**As authority is delegated down-line, it is diffused so that no one on the bottom level has the power base to overthrow the one at the top.** In fact, they cannot even get a call to the one at the top. In many of today's situations there are five layers of regional and underling oversight that prevent a person from contacting his pastor. When shepherds are deaf to the cry of the sheep, they are no longer shepherds. Pastor Greene could not even hear the sound of warning coming through authority greater than his.

**The problem with authority is that it has a reciprocal truth. Authority is answerable to responsibility. That's the tough part about pyramids. If you are ultimately in control of all authority, then you are also responsible for every good and bad thing done by every person in that pyramid. Responsibility comes crashing back up-line, but it is multiplied and is focused on one individual. The pyramid inverts and the pressure buries the one at the top.**

"Pastor is away for a few days, you know the pressure of the ministry."

"I wake up in a cold sweat, afraid the phone is going to ring and someone will have another problem."

"Since we signed the note for the bank, I have not been able to sleep."

"Yes he is very grumpy these days, but you do not understand the pressure he is under."

"Pastor cannot seem to shake these migraines."

Possibly the saddest one of these justifications I have heard is Billy Graham to Larry King in reference to the then President of the United States caught in promiscuity with a subordinate; "We have to take into consideration the pressures of the office."

Why do leaders at every level seek out prostitutes, steal money, and lie to themselves and family members? Why did Moses strike the Rock?

That's right, Moses struck the Rock. He hit the One who had saved him. He took up the miracle staff and smashed the provider. It added stripes to Christ. This was a spiritual rock that followed them in the wilderness. Where had the bitterness come from? What had caused the flood of anger which, unfettered, would cost Moses the Promised Land? It was the pyramid. This thing is a killer! (1 Corinthians 10:1-4).

Why had contention entered a group of men who in the peace and joy of retreat had covenanted together to serve the Lord? How had competition come in until home group attendance was no longer a joy but a labor? It all became clear. We were in a pyramid and, worse, I was playing the second fiddle game. "Who is Who" had given way to "What is What" and we were caught in the web of unbelief that complained about the provision of God. By the time we blew up, our very patient and loving Pastor had the rod in his hand and the Rock in his sights.

## A New Revelation

About this time, in 1980, I was invited by a computer programmer in the church to attend a demonstration by Hewlett Packard (HP). What, I asked myself, am I doing going to this thing? This is crazy. But the programmer was a good friend so I thought why not look into his world for awhile and, with joy, I went. What I saw there was going to reform my life and, as I look back on it, preserve me in the ministry.

HP was unveiling a new concept. They had found a way for the database to talk to the printer. In fact, you could integrate all of your systems so that you produced a document that could really communicate your concept to your audience. I was so very amazed at this that I just sat and stared.

That was it. If a Pastor or leader of any kind would step off the top of that pyramid and get into the middle of the people and build relationships, then they would be in one mind and one voice.

Within my sphere of influence I began to apply the concept. Get in the middle. Use the influence which God has given to enable others to rise to the task. Find the common voice in it all and express what the Lord is speaking. Find your success in the success of others. We survived the blow up of the pyramid and began to see growth. At this time I read a great book by John Haggai, *Lead On*.

He defines leadership as, **"The discipline of exerting special influence within a group to move the group toward goals of beneficial permanence that meet the group's real needs."** That was it. That was the word for which I had been seeking. "Within" is a wonderful word. Not "over", as is the manner of the Gentiles or Midianites, but "within". The way of Christ. The living way of the Word of God.

I began to see the joy of serving others in their vision. The success was wonderful. Communication flowed unhindered and from the ashes of dead dreams grew wonderful ministries that multiplied through the lives of the people to the nations. The Lord had provided the resurrection power so essential to multiplication.

I avoided pyramids like the plague. When a speaker would come and refer to "his ministry" a gate would close in my spirit. I tried to caution gently. Please, Brother, let it be the Lord's. In the anonymity of the hour I was able to become a servant to many missions ministries. We built a network of relationships that would stand through many tests. We did not have to own ministries to be involved in them, but became servants to the vision or others as long as it was a ministry focused on reaching the nations with the good news of Jesus.

## The Multitude of Counsel

Applying the principle of the multitude of counsel, I contacted four area Pastors to see if we could meet together Thursday mornings to pray. An Anglican, an Assemblies of God, a Baptist, and I began to pray for each other. The fellowship was wonderful. One of the first things we prayed for was the ability to be open about our feelings in situations. We were careful to not mention names in our conversations. Soon we looked forward to Thursdays. The breakfast became the primary meeting in our schedules. Eventually, the news of a good thing became contagious and there were as many as 50 gathered for prayer each week. Imagine, in a small town in Western Pennsylvania, 50 men of God traveled as much as 20 miles to be with a table of four who could help them leave the perch of the pyramid for the safety of the multitude.

With the numbers came the question of organization. The four of us felt very strongly that the Lord's calling us together was for nothing more than fellowship. In fact we made sure that the restaurant set only tables for four and that the breakfast was served family style rather than individually. I must say that we watched out for those who tried to "speak" rather than "listen".

When Pastor Greene got caught in that dark narrow side street he had not been with a group of his peers. In fact, he felt he had no peers. He was looking for intimacy, but the wells from which he was drinking could provide only the decay of Meribah.

Our salvation was in thinking of others as better than ourselves. I really admired the men with whom I met and prayed. I was very open with them about the pyramid in which I had been involved and the role I had played. They often prayed for me. We came into a bond of relationship that prevented side street derailment and kept us on the brightly-lit highway of holiness.

Together we explored the heart of the Lord in His declaration of a Body. We began to see the Body of Christ as larger than the total of its parts. We found that the Body is a system of related cells. While we differed in such foundational doctrine as water baptism, we agreed on love among us. Our points of view on tongues and the Pentecostal experience paled in the light of our need for each other. Even Baptist Calvinism joined with AoG Armenians. It was a day to behold and through it we were able to provide a safety net preventing many late night light ups.

*Chapter 3*

# Life Circles

JESUS has made for us a new and living way. The death I had tasted in the Pyramid was vomited out of my mouth through many times of prayer. I had been, as Isaiah, a man of unclean lips. I had joined in the battles between interest groups and those interested in ascending the pyramid. I had, as a result of ambition, hit the Rock. Instead of focusing on Jesus, I had fallen prey to the power of the Pyramid and the passion of People Pleasing.

If you are in the middle of one of these situations, please repent. You must live to please the Lord and you will become pleasing to those who also live to please him. You will be able to see Jesus in control of your life and not the organization of the local church. For his sake, you must get your pastor into proper perspective.

The Hewlett Packard luncheon had shown me a way out of the depths of dilemma. I needed to be a part of something that was integrated to serve one purpose, the Kingdom of God. Personal ambitions and the kingdoms of man had to pass away to make way for the King. We could no longer spend hours assessing the growth of the pyramid. We could no longer afford the time to consume our resources in building the great tomb of Egypt. We had to undergo a transformation from death to life, from Egypt to the Promised Land, from the tomb of temporality to the eternal throne room.

**Hebrews 10:5-7 makes the bold declaration that the will of God cannot be accomplished by the old system. "Lo it is written in the volume of the book ... a body thou hast prepared for me." Most certainly the direction for our lives and ministries must be found "in the volume ofThe Book."**

Jethro's pyramid and those which have followed are not in the volume ofThe Book. They are the way of the world. The church is not a staid financial institution amassing resources for its own consumption; it is a living organism developing resources to meet the needs of others. We who share the honor of serving her must have the same life-giving spirit she does. We must live for the development of the body, and not for the development of our personal agenda.

The "volume of The Book" speaks of life as Jesus boldly declares, "The thief has come to rob, kill, and destroy, but I have come that you might have life and have it more abundantly." Indeed, in Him is Life and that is life eternal. Certainly His body was given that He might carry sin to the cross and take it out of the way, but in the volume ofThe Book, He is the great life giver. It is He who has breathed upon dust and caused it to live.

How had Pastor Greene arrived in that dark, narrow, night of his life? He had left the protective warmth of the body of Christ and ventured out into the stark cold of the world. The loneliness of the pyramid had pressed him to pass the boundaries of righteousness and explore the power position of pyramid perching. As dark as the enemies whirling about him, he had sought a place where he did not have to ask, he did not have to relate, he did not have to serve, he just paid and was served.

Thank God the elders of the church had a positive pathway to bring him around this corner of his life and carry him back to the love and light of the body. They could have slammed the door. They could have left him in his hour of great need, but they were the body. They would never sit at the top. They would never be paid for their service to the church. They were just a group of tradesmen who voluntarily served Jesus and His church.

Simple in their approach, gentle in their ways, they opened a way of return to righteousness to a man whose pyramid had turned upside down. How were they going to do it?

## Restoration, Reaching out

**First, they came to him.** They did not make the pyramid power play of demanding that he crawl to them. They were not of that spirit. They met, formed a prayerful plan, and came to him with a lifeline. They reached out to their Pastor in his hour of need.

Life reaches out to those who have fallen. It finds a way of redemption. It does not condemn. Jesus came into the world not to condemn the world but that through Him the world might be saved. It is His ability to love at all personal cost that gives life to the church and hope for her future.

When a brother is caught up in a fault, you who are spiritual should seek to restore such a one in a spirit of meekness. Each one of us has areas which, if uncovered, would cause us to fall. It is the prayerful support of those around us that keeps us in the path of holiness. Pastor Greene had escaped or abandoned the covering of the church. He had gone outside. He had broken covenant and relationship. But, the elders were good to reach out to him and embrace him to restore him to the life flow of the body.

They had a choice. They were deeply disappointed. They were embarrassed, they were ashamed. He had caused them to be compromised in the community. They would face many uphill struggles as a result of his actions. The tremor caused by his indiscretion would reach through their families. He had been an example to their teens. He had spoken many words of direction and instruction to them all. Now his words would lose value. But, in the face of all this, they decided for life and embraced the man who had caused them such pain.

Trust = Love x Time

Never again would he be able to say, "Just trust me on this." Trust had been violated. **Trust is the demonstration of love multiplied by time, i.e. Trust = Love x Time.** Trust relationships are built over many years. Infatuation yields to commitment, which in turn gives way to dedication. The process is an experience of growth in selflessness or love.

Relationships in the body of Christ are trust relationships. They build over time. Each one is formed through selfless service. A person can be a believer and sit forever in church and never have to grow, but true discipleship requires us to grow together. We learn to trust one another through service. By virtue of his position, Pastor Greene did not have to serve. He had begun to be served.

## Life Circles

**Second, the elders placed him in the center of a small group of peers with whom he could be very frank concerning his personal life.** This was absolutely essential to his restoration. He needed to be in a "Life Circle." From the pinnacle of the pyramid his only way was down. Everyone drew from him. No matter who the person was in the church, his wife included, he was a "Priest to God for them." They were under him. He had no way to grow and reach up to the Lord. He could not ethically open up to another person in the church. As the Pastor, he was aware of information in the lives of those in the church. He had to carry this to the Lord, but he also had to shoulder the responsibility of caring for them.

This load had become too heavy. He sought relief in release that did not require relationship. Yes, Dr. Graham was right, we must understand the pressure, but we cannot see sin as the acceptable release. There is, in the life of the church, a place for strength. It is in the relationships of trust we form with those around us — in the "Life Circles" we form.

By requiring the Pastor to join with other ministers and begin the process of deprogramming the pyramid and coming back to life, these elders spared his spiritual life. They looked beyond their own resources and those of the church and found in the greater body a way of redemption. They were ready to restore the individual, and then address the church's needs. Such love is born in hours of prayer for their Pastor. It does not come easily. It is love demonstrated over time.

How had their love been demonstrated and developed to this point? In every check they had written to the church, every time they had cleared the walks or fixed the wiring, every time they had come early to open the building or stayed late to close it, they had demonstrated and developed their love. In a hundred quiet ways these men had served without thanks. They were devoted to the life of the body. They were caregivers, quiet servants who make the whole thing happen.

How do we know that? Look at their response in crisis. There was no long meeting, no attorney, no lawsuit, and no cover up.

Straightforward compassionate leadership is born of service. The man of words was broken; the men of quiet strength came to put him back together.

**They formed a "Life Circle" around their Pastor and his family.** They covered the man with official statements to the outside world and quietly addressed his sin in privacy. He had been removed from the pinnacle and, as he plummeted in his personal life, they made a safety net of compassionate concern. Within the cocoon of their protective love he could regain balance, redefine calling, and receive much-needed relief. He was looking for release in the night's darkness; he found love in the dawn's clear light.

*Chapter 4*

# Bridge Building

THE unity of this group of elders shows us that they had something the Pastor had not enjoyed. They had developed a close relational base for their ministry. Perhaps his absence from their lives had caused them to do this. Quietly, in a non-threatening way, they had been formed into a loving, body-serving "Life Circle." Drawn by their love for the Lord, His church, and their Pastor they had been prepared by the Holy Spirit for this moment of crisis. They even addressed material resources in their response to crisis.

## A Relational Base for Ministry

Building a relational base is essential to successful ministry. "Life Circles" are the basic unit of a strong relational base. Just as with the HP integrated systems, the "Life Circles" place the pastor or group leader in the role of servant to the other parts. He can, from this central position, distribute the resources necessary for the team to achieve common goals. The Pastor moves at the pace of the group and not his own. His success is defined in the success of the entire team. Weak members are supported by the strong and they are of one mind and voice.

In his letter to the church at Corinth, Paul outlines the body concept. He declares us to be members of one body, but we have different gifts and abilities. He is very clear that we are to stand together for the common good of the body and there are to be no divisions among us. He is also quite clear that the church will build itself up in love.

This love is demonstrated in one part complementing and supporting another. In Pastor Greene's hour of crisis, the church responded with love. That "Crisis Control" type of love must also be manifested in the daily building of relationships. The relationship that Pastor Greene was about to learn in his newly-appointed "Life Circle" is the missing part of many churches. It is that time when a Pastor leaves the pinnacle and gives himself to the members around him. It is what Christ did in the midst of the disciples.

*Bob had been a Christian for close to 20 years. In that time he had been part of six local churches. His career development had kept him moving from place to place. In every place he and his wife had found a local church in which to serve. Because of their transient nature they had never gotten past the "New Folks' Banquet" stage and were looking forward to retirement as a time when they could bond to a Pastor and become a viable part of the church. Along the way they had gained some experience in teaching and hosting small groups. Lesley was very good at one-on-one counseling with those young women who were facing the challenges of new marriages. Bob loved to fix anything that broke and they had left a trail of repairs through several states.*

They retired to a lovely small town near a major city and were very happy to see that a new church was emerging. Naturally, they came alongside the young Pastor and his family. They were eager for relationship, feeling that at last they could put down roots and develop a close personal relationship with the Pastor.

Bob was invited to serve as a member of the church board. During the interview with the nominating committee and Pastor he was given the distinct impression that they wanted to form a "Relational Team" that would carry the church to its destiny in reaching the community. Bob was very happy to be invited to be a part of what appeared to be a "Life Circle." To Bob this meant one thing — he was going to form a close personal relationship with the Pastor. This would be the first time he could get past a "new folks' banquet." In his mind it confirmed his move to this town, timing of retirement, and the leading of the Holy Spirit in his life. This was good news.

Excited, he rushed home from the meeting to share with Lesley the great news that he had been accepted in the body and even invited to be a part of the relational team that would form the core of the church. Against her instincts, he excitedly accepted the nomination and was affirmed by the church. This was it. All he had envisioned for the local church was going to come to pass. This was Jesus and the disciples moving from miracle to miracle as multitudes were saved. The lame would walk, the blind would see. He would grow in the knowledge of the Lord.

His pastor spent an overnight retreat with the group. Bob was thrilled. Bob liked the other men and there was a special feel to being part of the relational team. The fact that the Pastor even wanted to be called "Fred" was encouraging. Bob heard the testimonies of those around him and thought what a wonderful group the Lord had assembled. He was able to set lunch or dinner appointments with the others for the coming months. He could just feel the warmth and acceptance.

The thing that touched him most deeply was the openness of the group. Their prayer times had been rich and, as the Pastor shared his vision for the body, they with one voice had affirmed it. He was very excited. Returning home he reported to Lesley that her misgivings were unfounded and that this was going to be really great. He could not wait for the board meeting in two weeks to again be in fellowship with these men.

His first call to the Pastor that week was received with joy. Yes, they had a great time. Yes, there was a certain peaceful presence of God throughout the retreat. Yes, this was a good time in the life of the church. No, he would not be able to make it to lunch, perhaps a rain check. Sure, please do call again.

Later in the week Bob called just to pass on a part of Scripture that he felt the Lord had quickened to him during his morning prayer for the church. The secretary assured him that the Pastor wanted to hear from him, but he was very busy just then and could he call back later. Bob did, but the Pastor had gone out for the day. Lesley didn't say much. Her silence said it all.

No further retreats were scheduled. Board meetings began with prayer but soon degenerated into the same fractious affairs Bob had heard about in the past. Bob approached the Pastor about his feeling that this was far from the "Relational Team" about which he had been so excited. After six months the joy was gone. Bob finished his term on the "Relational Team" and began to host a weekly Bible study for people from the community.

## What Had Happened?

I spent a lot of time with Bob and with his Pastor. I listened to both men and found they told the same story in different ways.

Both said that they wanted their relationship to grow beyond what it was. They were both honestly interested in developing a close relationship. They realized that their differences could become serious problems in the life of the church and neither of them wanted that. They were willing to work on it.

The final blow for Bob had been an instance in which the pastor had told him one thing and a staff member another. Bob had felt that the Pastor had been dishonest in handling the situation and had gone to the man to express his point of view. Pastor had become defensive, irritated, and had said, "I am the Pastor here and if you cannot live with that, perhaps you should find another church."

Does that sound like hitting the rock to you?

*Bob completed his term, and did not seek to be on the team any longer. He and Lesley remain in the church and serve with helping new believers and doing follow up as they always had before. They continue to fix whatever breaks, but the relationship they had so strongly desired has passed them by.*

Lesley says, "Hey, after 20 years, what did you expect?"

*The Pastor expresses relief that Bob is not a part of the board. He feels that Bob is unyielding and that Lesley has a sharp tongue. He says they are not "Board material."*

## One Body

The church is the product of its members. Each one of us brings to the whole our understanding of Christ and His love. Each of us has gifts, which are supernatural, and abilities, which though God-given, are quite natural. These abilities fall into three basic categories: head, hand, and heart.

People are generally strong in one, average in another, and a disaster in the third. You would never want me to try to finish your wallpaper. I have the hand skills of a walrus. On the other hand, if you had a deep felt emotional or relational need and called upon me, you would find a heart that is ready to embrace your joy and sorrow, and walk with you through anything. I need some help with my head. I have to hear something several times from various points of view before I can understand it.

Bob can fix wires. He knows what goes where and how to make it all happen so that when I find the switch to turn it on, the lights work. His Pastor is a brilliant man. Deep in the Word and understanding, he sees the complexities of situations that Bob in his "lights on" or "lights off" thinking does not see. Bob's world is black and white; his Pastor lives in shades of gray.

In the body we are all of different gifts and abilities. No two of us are exactly the same. We can choose to differ over this or we can realize that the body is made strong by the maintenance of a healthy tension along the joint.

## Strength in Diversity

One of the men in our church owns a fitness center. He is the perfect specimen of physical excellence. He works out every day. In his 30s, he still draws the attention of everyone when he walks into a room. Even his forehead has muscles and they are trim.

When he lifts weights, he uses two muscle groups that oppose each other. Tension along the joint of his elbow balances the pressure and his muscles grow and are trim. His muscles actually oppose each other during the work out. When he uses that arm to do work, the strength developed in the joint tension provides the power necessary to do work. Paul says that the body is fitly joined together by that which every joint supplies. What is that? That is tension or the celebration of diversity. We are not all the same and we are not all pulling in the same direction.

What does this have to do with Bob and his Pastor? Everything, they are simply out of joint. Neither of them wants to divide the relational group of the local church, but for the moment their efforts are across the joint. As long as they maintain the integrity of the body, and respect each other's gifts and abilities, they will grow stronger in the process so that when the body reaches out to the community they will be able to sustain together the work of the ministry.

Though there is a strain on the relationship now, it will become strength as they pull together for a common purpose. Selflessness demonstrated over time will heal the disappointment of the previous offence. The multiplication factor of (Trust = Love x Time) will again work as soon as they demonstrate forgiveness.

## A Living Example

**Here is the key that unlocks the fulfillment of the "Relational Group" from the chains of the "Board Material."** How did a carpenter bring together fishermen and a tax collector and mold them into a ministry base the extension of which would reach around the globe, create history, direct nations, and reach the limits of space and seas? He lived with them. There was no professional distance in the ministry of Jesus. He taught them a lifestyle by living it before them. Jesus made Himself available to those He wished to disciple.

When Andrew and John left the ministry of John the Baptist to follow Christ, Jesus asked them, "What do you want?"

He knew that the way to realizing His desire for a vibrant group of believers who He could fill with His nature and power, enabling them to reach every man, woman and child, was to first serve them. He set himself in the midst of a multi-ethnic, multinational, multilingual group as diverse as any has ever been. Galileans who could not write or speak the common trade language were matched with the eloquent of the business community well versed in Hebrew, Greek, and Latin as well as their mother tongues.

He did not say, "Come, build with me."

He did not say, "We can make a difference."

He did not say, "Come and be saved."

He said, "What do you want?"

They wanted, according to John 1:38, to know where and how He lived. They saw in him a way of living, different from any they had seen. He was the way, the truth, and the life. They could sense a difference in this man who had filled their nets. He was not the distant leader that was so prevalent in those days. He was touched by their infirmities. He was one who would live with and die for them; eating the same food, sleeping beside the same fire, riding in the same boat, staying in the same places. He had no place to lay His head, but He would make an abiding place for them in eternity.

**They wanted a leader who would identify with them, as they were, One who would not rise above them, One who would speak their language and share food with them. They were looking for someone to walk with them in the trials of life. Jesus did that. He joined Himself to them. He became one of them that they might become one with Him.**

He stepped from eternity into the midst of humanity and became the servant of all. Lord and servant in the same package. The fisherman could now talk with the tax collector. He created a "Life Circle" of integrated living stones, a mosaic that showed forth His selfless love. That "Life Circle" would stand in the face of great persecution and demonstrate His love. Those who joined that circle would later be identified as those who had spent time with Him. This was the mark of their fellowship — they had spent time together.

This is what Bob had wanted. This is what Pastor Greene's elders had wanted. This is what the two Pastors coming to his side were going to provide. This is the living, life-giving way Jesus made for us. Each of us, unique in gifts and abilities, can be forged together in the heat of His love to present His purpose to the world that they might believe.

**But, you ask, how can I make the change? Will it work in my family? Will it work in my business? Yes, and as we press on together we will build the BRIDGE that will take you from "A" to "B".**

*Chapter 5*

# Trust = Love x Time

**TO build the BRIDGE that will take the body from the turmoil of Pastor Greene and Bob to the fulfilling joy of Peter, James and John we must first build a strong relational base.** This is not difficult if the pain of the disappointments in pyramids is allowed to be healed, by restoring the **"Trust = Love x Time"** formula through a "Life Circle." We can learn from Moses' example. We can gain insight from Jethro as to the source of contention. We can see from Paul's writings the model to follow. And we can call upon Jesus who dwelt in the midst of that first relational group to form through us "Life Circles" that will demonstrate to the world the love of Jesus Christ. "By this all will know that you are my disciples, if you have love one for another" (John 13:35). Let us lay aside the bondage of Egypt and move on to the new and living way, which is in Christ.

Lee Iaccoca was a brilliant project director. His vision for the Ford Mustang and the Plymouth Voyager created employment and provision for thousands of workers and provided practical and economic transportation for millions of consumers. Yet, it was his ability to manage innovation that speaks to me. He believed in the team concept. Form a strategy group and allow the synergism to carry the vision to fulfillment.

## "Life Circles" in Practice

Mr. Iaccoca taught that each of us has only enough strength of character to handle five portfolios. Think about the Pastor as the CEO of a small corporation. How many portfolios can he handle? If he prioritizes his personal walk with the Lord, he has only four other spaces to fill. He can attempt to divide the church into four sections. Let's say he tries 1. Worship; 2.Visitation; 3. Education; and 4. Discipleship. Good enough, he will succeed in four aspects of ministry.

Let's say he approaches it a little differently. What if he chose 1. The Garcias; 2. The Chans; 3. The Gianuccis; and 4. The Smiths? Let's say that each of those families embraces four families. Now he is successful in extending his influence along relational lines to 21 families. As these families are encouraged to identify the areas of ministry in which they will be involved, the whole body will be fit together through that which every joint supplies. He has shifted his paradigm from directing projects to pastoring people. He will be happier, they will be happier, and the body fitly joined together will grow up into the Head, which is Christ.

He still has the ability to develop personally. He has family time. He has good Word preparation time and he has covering accountability with these four. He has formed a "Life Circle" with these families so that communication and interaction is involving more people than ever before. If one of the families gets "out of joint" the "Life Circle" can embrace the situation. The body can heal itself. He has only to have one part communicate to the rest.

There are three risks for him. He will not be able to be everything to everyone. He will appear to the untrained eye to be partial in his love. His ministry will be accomplished with such little strain that the uninformed will view him as a bit casual.

## Relationships of Trust

The "Life Circles" are sustained through relationships of trust. **Trust is Love multiplied by Time.** The longer a time you demonstrate love or selflessness to those around you, the greater a bond of trust is established. Peter, James, and John were looking for someone they could trust. Bob was looking for a leader he could trust. Lisa Greene was reaching for the shoulder of a man who had violated her trust, but she was saying, in doing so, "You have loved me for a long time, not all trust is gone. I am willing to start today to restore that trust."

The elders were saying, "It will take a year to restore the breach of trust that has occurred". Love reinstituted and multiplied by time will result in trust being restored. But, if love is not demonstrated, or time not allowed, trust cannot be maintained. Any value multiplied by zero will result in zero. Remove love or time from the equation and trust cannot be regained or sustained.

In one of our Asian cities, I have a very strong "Life Circle" established. One of the brothers there has walked with me through many nations. Through many dangers, toils, and snares, we have already come. Recently I was asked why we are so close. "Easy," I answered. If he says that on a certain day at a certain time he will meet me in a certain city of the earth, then I know he will be there.

If I get there and he is not there, I know he is on his way. If he gets there and I am not there, he knows I am on my way. Our word is our bond. Selflessness over time has created trust in our relationship.

Due to the nature of the work I do and the pressures of raising a family, my wife and I are apart more than half the year. How can our marriage of 30 years be prospering? We trust each other. Selflessness over that time has produced trust in our relationship. It is not the formal declaration of marriage, the license, or the disaster of divorce that keeps us married, it is the trust that has developed. Love, given time to operate, will never fail.

My son recently asked why there is so much divorce today. While there is no easy answer to this question, one of the reasons that I feel most strongly affects the marriage relationship is premarital sex. Couples run to pleasure before they run to trust. It is no wonder they do not have trust at the core of their relationship. They were self-serving in the formation of it. Relationships built on trust are life giving. They do not demand; they do not step outside of respect.

So it is in the intimacy of relationship to ministry. Say you want to begin an outreach to the young people in your community. You can build a pyramid in which you are the star and others give to your ministry, or you can build a relational base, which will last for several ministry seasons. It is through your relationship to the others in your "Life Circles" that you will reach the young people around you.

In his letter to the Ephesians Paul teaches that there are differing gifts which have been given to men so that something called the church may evolve. As these gifts are brought together under one head, which is Christ, we see the strength of the body prepared for Him. We pull together through that which each joint supplies until we reach a mature body, one that is able to reproduce.

**We begin to realize that in the local church the Pastoral gift is central and the others peripheral.** That is to say the pastor relates in close, personal, trusting relationships to four different types of people. He demonstrates impartial love and service to see the growth of the ministries of apostolic, evangelistic, prophetic, and teaching people. He will live and die with them. He will love them over time, laying aside his personal ambitions that they might be established in ministry. This is what Jesus did, though He was God. He laid aside His own things to serve the needs of others. This is the Christian ministry.

## The Apostolic Ministry

Apostolic people see only the mission fields of life. They measure all that they are and do by the hardships they face and the enormity of the task before them. They are driven people. Fiercely independent they can become a strong divisive element in any ministry. No matter what the sacrifices of those around them, they radiate an air of superiority that drives people from them.

Apostolic people are so concept oriented that they forget to return calls. They can be very rude. In a ministry team they will exhaust the others with tales of adventure in foreign lands or alienate others with a great deal of self-serving rhetoric. This is the nature of the gift they carry. They often do face the impossible on a daily basis and have learned that it will be Jesus and them alone in the face of danger.

True apostolic ministry has paid prices, has learned a second language, and has lived through hardships. It has seen the Lord perform the miraculous. It is at times intolerant of the seemingly slow pace of the teaching ministry, or the mystical approaches of the prophetic.

Apostolic people move about in an "is" or "is not" world. Something either "is" or "is not" Christian. In this way they agree with the Prophetic, and have little appreciation for the Pastoral, Evangelistic, or Teaching "work in progress" ideas. In many ways they are like Bob in his "lights on" or "lights off" world of black and white.

When one of these arrives in the life of a local church, they must be embraced rapidly by the Pastor. A "Life Circle" is the perfect place for one of these returned missionary people. They can form healing relational bonds with people who are not from the same point of view as they. A Pastor living at the top of a pyramid will push these people away as they pose a real threat to his position. A Pastor in a "Life Circle" will embrace them, as he has no position to defend.

A man who is hungry for real relationships pastors one of the churches I serve in the US. There are now six returned missionaries in his congregation. Each one of them has integrated successfully into the life of the church by being embraced into one or another of the "Life Circles" that exist. Their resultant growth in Christ is wonderful and no shock to those around. Others in those "circles" are beginning to go out on short-term trips. One young man has responded to a call to mission and all of this in less than five years.

## The Prophetic Ministry

Prophetic people can be equally challenging. Everything in their world revolves around their perception of the will of God or the Word of the Lord. Their mystical approach drives the practical evangelists and apostolic people crazy. In a meeting they just want to sit in the presence of God and wait upon Him. Good ideas must give way to their "inspired" point of view. To relate to them, a Pastor must be able to embrace their mystical airs.

There comes with these folks a sense of impending action. God is always about to do something. For the practical Pastor this can be a bit nerve wracking. There would, at surface value, appear to be disdain or disinterest in the wonderful things the Lord is doing. And, in its place a certain obsession with the future and lack of thanks for the present.

The pyramid Pastor will be very nervous around these folk. He can never be certain when they are going to receive some insight from the Holy Spirit concerning him and, if they feel led, they will just bring it right out. Pastors in "Life Circles" are secure in their relationships with these folk to embrace them as a gift to the church. They form a trusted safeguard against straying from the presence and passion of Christ.

No ministry base is complete without the prophetic. Through this gift we see our part in the greater whole of the purpose of the Lord in His calling. As the Apostles and Prophets ministered to the Lord and fasted, the Lord gave the direction of separating men to ministry and expanded the church across the earth.

The prophetic flows from praise and worship. These gifts are essential to the development of any ministry base. If your BRIDGE is going to reach the lost, it will have to have a position of praise. Wise is the teacher who realizes that the "wait" until worship is over is the most important "delay" of his day. It is the anointing of worship that clears the air for the proclamation of the Word of God.

"Souls, We must reach souls!" the cry of the Evangelist is always followed by, "Follow up, who is going to do the follow up?" of the Teacher. I have spent hours in the middle of this discussion. In the pyramid, the Pastor delegates follow up to staff. In "Life Circles" each is responsible for his own. This appears to diminish the ministry of the evangelist, but not so, for as each circle grows and multiplies, it is this gift that compels the whole body to the heart of God for the lost. It is this gift that says to the others, "Let us never lose sight of the real reason we are here."

Once a Bible College student burst into my office to declare, "I will not sit another day in that brother's class. I cannot stay awake. He makes no sense to me. His teaching is so very dry, I cannot stand it!"

"How," I responded, "Can the book of Romans be so very dry to you? It is the gem in the ring of the epistles."

He just looked at me in disbelief. I, the apostolic, had just quoted the dry old teacher. I had affirmed my brother from the "Life Circle" that was molding the kingdom in the life of this young man. Now a very successful missionary, he loves to teach the book of Romans and whom does he quote most often? Of course, he quotes the teacher.

The power that was transforming this young man was the relational trust that had been forged in the lives of those of us who had seen the pyramid come down. In the face of ruin we had declared ourselves to each other and the fruit of that "Life Circle" has reached the nations. Jesus said, "By this all will know that you are my disciples, if you have love one for another" (John 13:35). He formed a "Life Circle." Forsaking fame, He gave Himself to weaker men that His life might flow through them.

**To build a bridge from "A" to "B" you must build a relational base. It does not have to be large. Start with four**

**others. Place yourself in the center as a server to them.** Carry the life flow of the Word of God to their giftedness. Pray for them several times throughout the day. Be mindful of them and sow into their success. Know their goals and move resource toward them. You will be amazed at their joy. You will rejoice with them in their triumphs and walk with them in their trials. Together, your group will accomplish what none of you could do individually.

*Chapter 6*

# Spiritual Resource Development

"AS of this date, I am resigning as the Senior Pastor of this church. The differences that have formed between the pastoral team and the elders appear to be irreconcilable, and in the love that I feel for the church and after counseling with close friends in other ministries, I feel that this is the action I must take."

Do you know what this means in clear speech?

We are broke. Not broken, broke. We have no money. We cannot pay the Pastoral Team. There is no cash. This is independent church speak for, "No money."

When I received this letter I was deeply moved. Pastor Jon is a friend of mine. He has been very good to me. We have known each other for years. What was he doing at the top of or trying to get to the top of a pyramid? The two friends to whom he refers are both pyramid dwellers. Of course they would tell him to create a public power play by circulating a letter to stir the fires of loyalty. The letter left me heartsick.

Why does he not have close relationships within the eldership? Why was he turning to men outside to resolve issues inside the ministry? Because, "Life is lonely at the top."

I did a little research into the situation and found that the attendance had fallen to a point that could not sustain the budget of the work. This is a resource crisis. Staff can no longer be paid; obligations that have to be met will be fulfilled through another loan against the value of the facility.

Now the search is on for a new Pastor. Now the whole church re-evaluates their relationships. Will the elder's "Life Circle" hold together? Not this time; they are already leaving for a more "loving" and "stable" situation. The outreaches of the church, food for the hungry, homes for the homeless, foreign mission commitments, all are in danger of closing down. The last stones of the pyramid are about to seal its contained corps for all time. I am sick about it.

## Developing Resources

**A successful ministry meets its goals in the time allowed within available resources. Resource development is one of the great challenges to ministry. This is the "R" in our acrostic and the second building block of a successful BRIDGE to take you from "A" to "B".**

Resource flows to vision. It is the responsibility of your "Life Circle" to develop and declare your vision. You should apply time in the prayer and fellowship to completely explore the vision the Lord is giving you. All five gifts should interface to express that vision.

Many years ago I was hired to develop a church-based bible-training center. This involved interface between an existing night school, a Christian Day School, a Pastoral Staff, and an active Eldership in a rapidly growing church. After researching the state of such schools in the US, I decided to form a "Life Circle" whose primary function would be prayer.

All success comes from prayer. Your vision must be conceived in prayer, birthed in prayer, bathed in prayer, sustained by prayer, matured in prayer and, in this case, buried with a prayer. We met weekly for several months and sought the Lord together to know His will. At each meeting I introduced a topic such as "Curriculum" and we prayed, asking the Lord to show us how to develop the curriculum. The "Circle" was composed of ten men and women from a cross-section of the church. I supplied a written report to the pastoral staff after each meeting.

## The Importance of Prayer

The Lord had led us into a wonderfully simple truth: Spiritual Resources must be developed before Human and Material Resources can be developed. The successful BRIDGE is one in which there is a vital resource development program. The first step to this success is Spiritual Resources Development.

Spiritual resources are the prayer, praise, and worship components of your vision. E.M. Bounds declares that prayer brings the power of God into every situation. Prayer is the slender nerve that moves the muscle in the mighty arm of God. Through prayer all things are possible and without it, nothing will succeed.

At the time, Kenneth Hegan Ministries was growing and there was a charismatic migration to Tulsa. I can recall Bro. Hegan saying, "Do not pray to get God to bless your vision, go to Him and get His vision, it comes blessed." With this direction we set our hearts to seek the Lord and, over several weeks, we were able to agree on a statement of vision, 'An environment in which the Holy Spirit can form people for His service'."

That vision statement led us away from argument over doctrine and into the creative energy of environmental development. It came through prayer. Because we put prayer first, we were able to enter into the creative capacity of the Lord and avoid the hierarchy of human understanding. Our meetings were 90 minutes in length and generally divided into three, not always equal, parts.

**First, we reviewed what the Lord had shown us the week before.** This report was given by one of the group who we had appointed as a scribe for the purpose. As I read this account, I was often amazed at things I had missed in the meeting. Seeing the vision develop through the eyes of a third person was wonderful, as I was free from the limits of my own understanding.

**Next, we began to thank the Lord for parts of the report.** That is to say that we would take turns praying with thanksgiving and then enter into worship.

**Finally, we spent an hour or so in prayer and worship with the scribe recording any directions we felt from the Lord.** Because we were open to Him, He brought Scripture to mind, spoke prophetic words through group members, led us in direction in prayer, and sent a deep peace to accompany those thoughts with which we were all in agreement. Spiritual resource was being applied to practical problems and we were learning that prayer is the most practical thing you can do.

That meeting became the high point of the week for us. From that room we could see graduates going into all the earth. We dealt with problems that we would not face naturally for years, but in prayer we sealed the Lord's victory. We gained knowledge and wisdom in His ways that would be applied more than 15 years later. There was a great sense of working together with Him.

When Bob was asked to be a part of a Relational Team that would serve at the center of the church, this is what he thought he was being invited to be a part of. Not unlike the times Jesus invited the disciples to join with him in prayer. Each man in the body wants to be able to be a creative part of that body. They respond to invitations to release spiritual resource for through that release they will be a part of the vision of the church. Bob thought this was the invitation. How great was the disappointment when he found it was just more meetings with the minds of men ruling and a pyramid being built.

In this "Life Circle", there was no title or position. Jesus was on the throne and we were very thankful to gather at His feet and hear from His desire for the work. Instead of Moses on the mountain and Joshua and Aaron below, we were together in His presence receiving the life-giving word that would bring us into promise and many thousands into salvation.

The prayer bug caught on and soon the whole body was infected. Each ministry began to adopt the "Life Circle" concept and began to be an exciting place to be. We developed into quarterly all-night prayer meetings where each of the staff would bring the folders of his areas of responsibility and open them to the people and we would all pray line by line through the life of the church. Everyone was included. The Lord was invited to direct by His spirit. The people could watch the vision received in those times unfold before them. They became the living body of the vision that had been released in prayer.

Small groups formed around vision and not need. We all were of the opinion that until the ministry was conceived in prayer, it could not be brought forth. Good ideas were referred to the "Life Circle" and were bathed in prayer. Prayer became the theme of the entire work. We were known for our worship and outreaches, but the real power of the place was its prayer life. The phrase, "I will pray about it" became much more than a convenient brush off. It had substance.

Bible School students came to the building at 6:30 AM to pray for 90 minutes before chapel. They filled the altar, walked the halls, stopped at the door of each Pastor and staff member and prayed for them. Miracles happened, not just in healing and deliverance, but in administration as well. Resources were being developed. We were learning as a body that prayer is the essential resource for the life of the body of Christ and that "Life Circles" would form around that pulse of prayer.

Those who were in the body at that time became ambassadors of prayer. They have now gone to six continents and over 150 nations. The deposit made in their lives of the spiritual resources developed in that initial "Life Circle" has now multiplied into prayer movements. Two of the men on the pastoral team in those days have traveled the globe bringing a prayer message. The students are now living in India, Thailand, Indonesia, China, Africa, Latin America, and Europe. Their ministries are touching thousands because they learned to keep spiritual resource at the forefront and allow all other resource to follow after.

**A financial resource crisis like the one faced by my friend is the symptom of spiritual resource erosion.** Why were the differences mentioned in his letter seen as irreconcilable? Because they were the expression of the points of view of two people who had their own agenda and had abandoned the Lord's vision. They were not willing to pay the price in prayer to regain it. My friend had climbed to the official position of "Sr. Pastor" to fire his last volley. The others will hold fast to their position somewhere beneath him and the cannonball will rebound in his face. Responsibility for the situation will be forever seen as his, even though all parties were involved.

## Practical Applications

**Spiritual resource is that combination of praise and prayer that fills the heart with the ability to believe God for His promises and the mind with the peace to wait upon Him.** It declares that the Lord is the head of the church and not man. That it is his body and not ours. Certainly we are members and each of us has a function to perform in relation to all the others, but he must be given the fruit of our lips and hearts in everlasting praise before His throne. Development of a spiritual resource plan for our lives is essential to ministry success. Here are a few ideas:

- Ask ten people to pray for you by name each day.
- Find ten people for whom you can pray each day.
- Set a fixed time in your daily routine to pray for others.
- Purchase gifts for your ten that will draw them to prayer. We have used the YWAM Prayer Journal and the Upper Room Devotional guides. Books by Andrew Murray, E.M. Bounds, Joy Dawson, Peter and Doris Wagner and Dick Eastman have proven to be good tools to motivate in prayer.
- Use the *Change the World School of Prayer*, or *Journey to Power* series as a small group developer. We did this in one church and their intercessors now span the globe calling on the Lord. Cities are changing because of this one "Life Circle."
- When the weather is good, take people out to prayer walk the community. Ask the Lord to give you a place to walk and pray. You do not have to be many in number to be effective.
- Begin a prayer journal. In one column write the date you prayed and, when the answer comes, record it. The Lord does answer believing prayer. Today's answer builds tomorrow's confidence.
- Begin weekly Morning Prayer sessions at your church.

- Begin prayer before the services.
- Ask your Pastor if you can host a quarterly all-night prayer meeting.
- Start a phone or email prayer chain to involve others in the prayer life of the body.
- Circulate a prayer letter in the church with quotes on prayer and needs-and-answers columns.

Jesus said to pray to the Lord of the Harvest that He would send forth laborers into the harvest. Human resource development must be a function of spiritual resource development. Pray in your workers. You want the ones He will send.

*Chapter 7*

# Human Resource Development

PASTOR Naomi Dowdy leaned across the table. Her features were radiant with the glow of thanksgiving. "Mark, you would not believe what the Lord has done now." I love getting together with her. She is always declaring what the Lord is doing now. She is so very excited about Jesus that you should know that she has hundreds of prayer warriors praying every day for her.

"What," I asked with a smile, "has the Lord done now?"

Sitting back in my chair I expected the usual testimony of a great healing or some awe-inspiring tale of deliverance or divine provision.

"He has brought us people with expertise in business." She was as happy as the day she told me of a service for four hundred, or an all-night prayer meeting. She was thrilled that the Lord had brought into the church she serves people with expertise in business.

"Where did they come from?" I asked, thinking that there was a great strategy about to be revealed.

"They were there all the time," she asserted. "Right in the church, but I didn't know how to release them into the gifts and callings they have. I thought we should convert them all to preachers, but they do not preach. They do business. And they are good at it.

"For so long we have treated our business people as though their 'secular' employment was in some way dirty. We have missed it. These people know how to develop a vision statement. They know how to resource a vision. They know how to make and handle money, some of them in large amounts. They know the market place.

"We have asked them to lay aside all this gifting to be Christian, but we have missed it. They know how to present Christ in the marketplace. They live Christ in the midst of the nonbeliever. They are powerful in their witness.

"All they needed was to be affirmed and released with purpose. They were there all the time."

Her excitement was contagious. It spread through the coffee shop as people shifted in their chairs. Looking beyond her I could see several suited folk listening in. When she gets excited about something Naomi can fill a room with a whisper. More than half the inhabitants of the five-star restaurant were now tuned in. The two brothers sitting with us had folded their hands in their laps and their faces were set in rapt silence basking in the glow.

"We have discovered several key businessmen in the congregation. They understand leadership, they manage resources, and they can formulate and implement strategy. All they needed was a little input and assurance and they have become a witness in the marketplace. Relationship with them is a two-way street. We can equip them as a Human Resource to reach those whose lives we are not touching. When they are accepted and affirmed, they assess the marketplace and make a lasting impact."

I watched her face. Her joy was in the development of the human resource. Her interest is not the money. She has handled many times the thousands this would make. The properties of the church are in the multiple millions of dollars. She was excited because the people she serves as Pastor had found a way, in the context of their vocations, to present Christ to people in the context of their real need. Naomi is a master human resource developer. She understands that, after the Holy Spirit, the people are the greatest resource the church has.

## "Life Circle" Principles

Human resource development is a great challenge to many of us. Two dynamics are involved. First is the balance of leadership and manipulation. "Find me some people to clean the church on Saturday" is a call to manipulation. "Teach the people stewardship and development of the gift God has given us" is a sign of leadership.

John Haggai writes that leadership is the discipline of exerting special influence within a group to move the group toward goals of beneficial permanence that meet the group's real needs. By teaching from within the group we establish three very important "Life Circle" Principles.

## Leadership From Within

In the pyramid, everyone is over someone. In "Life Circles," the leader leaves the position to which he is entitled and joins the group at their level. The abandonment of the "Father/Provider" role to assume the "Brother/Coworker" role exposes the leader to the possibility of rejection because the mystique of position is removed. As Moses' anger increased and anointing decreased on his path to rock striking, he had to veil his face before the people. Imagine how different Pastor Greene's outcome could have been had he not veiled his face before the elders who would eventually cover him.

This is exactly what Jesus did as reported by another great leader, the Apostle Paul. In the letter to the Philippians, Paul tells us that five evidences of the nature of Christ were seen in His leaving the "Over" position for the first "Life Circle." These will become the breath in our "Life Circles."

**"Therefore if there is any consolation in Christ, if any comfort of love, if any fellowship of the spirit, if any affection, and mercy, ... Let this mind be in you which was also in Christ Jesus ... made Himself of no reputation ... humbled Himself ..." (Philippians 2:1,5,7,8).**

When a leader has the same experience in which Pastor Dowdy is rejoicing, he will see the value of the people around him and center himself in the group to give them what they need to be successful in that which the Lord has given them to do. She becomes a voice in the midst. Five words from the text above will come alive in that fellowship.

**Consolation** is the translation of the Greek "Parakletos," which literally means "to call to one's side." It is the verb form of the noun, which we translate as the Holy Spirit. He is the One who calls us to the side of Christ. The function then is very clear. In a "Life Circle" we are not called to "Rise Up" to a level of seeming spirituality, but in the true nature of the Holy Spirit, we are called to "come alongside" those in the body. Instead of making preachers out of the business people in her church, Pastor is coming alongside them to walk with them in their vision. This action will result in their moving toward goals of beneficial permanence, which will meet their real needs. This is leadership, Christ style.

**Comfort of love** is the translation of the Greek word "Paramutheon" which literally means, "to speak to in a friendly fashion." Was it not Christ who said, "I no longer call you servants but friends for the servant doesn't know what his master is doing." Pyramid groups never let the membership know the full picture. Followers ask for the vision, they try to serve out of loyalty or a sense of devotion, but when the leader goes on a mountain top retreat, they form idols. Why? Because they are not spoken to in a friendly fashion. This is the relational group to which many feel they are being invited. They want to be the friend of the Pastor or team leader. They want to know the vision, the big picture.

All resource flows to vision. Human resource runs to it. Without a vision the people perish. They wander about in the vise of loyalty and ignorance. Feeling like fools, men who lead corporations find themselves following blindly after someone who veils his face. Is it any wonder that in America 80 percent of those who say they believe in Jesus do not support a local church?

## Relationships of the Spirit

The third essence of "Life Circles" is the **Fellowship of the Spirit**. Here the Greek is most clear, "Koinonia Pneumates." When title is set aside and hearts are revealed in trust, love multiples over time. Relationships pass from peripheral through practical to personal and then to perpetual, as we find true Christ love for each other. Fellowship in the Spirit is born in the heart and communicated to the head, causing the release of the hands to benefit each member of the "Life Circle." Just as the Disciples moved as one at Pentecost, so the "Life Circle" will ripple throughout the world with the good news of Jesus.

So much of church life is peripheral. People are known through the directory and may attend a new member's class and make four friends, but greeting is rarely by name and often we walk right past each other not even realizing that we are supposed to be parts of the same wonderful body of Christ. It is often very difficult for new believers to make it through the first two years of church life to become functioning members of the core or the "inner circle" or the "church within the church."

Bob and Lesley have passed through 20 years of this because of having to relocate in their professions. As a result, they have reacted to the latest "rejection" by forming a group to follow up new converts. They will be held suspect by the pyramid dwellers as a "non-related group" but the truth is, they will bear more fruit for the kingdom in the long run.

Our usual solution is to move people into practical relationships. These center upon ministry tasks. Ushering is a big favorite because you get to learn the names of the other people in the church. Nursery is not usually a good practical group to get into because those who are going to leave their bundle with you look at you with a very discerning and critical eye. You have to pass a police check to be in there. It is messy. And it is the one place in the church where you hope everyone in the room sleeps the whole way through the message.

Choir is fun. You get to sing praises, you face the crowd so it is easy to see and be seen and, after you sing the anthem, you get to let your mind continue in the heavenly while the message is on. It means a night out a week with people who like to do the things you like. But watch out for the holidays.

And so it goes, human resource filtering into fellowship based on the practice of ministry life and not the fellowship of the Holy Spirit.

One dear saint who gave her life to others in a circle used to say, "That's it! These people are turkeys! I am taking my fish off the door. There will be no more late night meetings at my house." Personal relationships come at a cost. The group is together for the purpose of discovering Christ in our midst and believing together for each other's well-being.

Personal relationships are trust relationships. Love or selflessness must be manifest over time for the true fellowship of the Spirit to be realized. We can live years in practical relationships and never find out what causes a brother joy or sorrow. We can stand next to each other as ushers or in the choir and never come to affection and mercy.

It is when we pull away from the pyramid and into the transparency of "Life Circles" that we learn to be merciful for we realize the amount of mercy which must be flowing for others to tolerate us. Imagine if Pastor Greene and his elders were in the custom of eating out once a month with their wives and one of the brothers noticed that the Pastor was a member of the "Second Look Club." That is to say that every woman who walked by the table received a second glance from his eyes. Wouldn't that observation have given the man the opportunity to minister at another time to his brother?

Close fellowship born of the Holy Spirit gives us the covering we need in a world that is set to destroy every Christian. Love does not count sin. Love covers, it is not offended. It does not seek its own. It is not lifted up. Love seeks the benefit of the others in the group, taking no accounting of personal cost, for the benefit of the group is the objective of the fellowship of the Spirit.

Perpetual relationships have love at the heart. Twenty-five years ago I was a young member of a "Life Circle." To this day I am mindful of those with whom we related. Some have already gone to heaven, but most remain. I think of them in kind and good ways because our relationships were born of the Holy Spirit. Pray today that the Lord will provide for you several very close, Spirit-born perpetual relationships — those that will carry you through the tests of time to eternity.

How do you know when it is happening? **Affection**, the fourth essential will begin to flow in giving and service. You will not need a duty roster to tell you that your brother has a need. You will not look to a program to tell you how to help another. You will not need a stirring message to pray for one another because these four and fifth essences of "Life Circles" will start to flow naturally. You will have moved off the pyramid of your mind and into the center of a group of people who you love and are ready to forgive.

**Mercy**, the fifth word becomes your motive in all things. The highway of redemption is paved over the old road of rejection as mercy makes each man your brother and each woman your very precious sister. No longer do we know one another after the flesh. We who are born of the Spirit begin to relate after the Spirit. A man is no longer involved in secular business; he is a brother for whom I pray that he prosper in reaching souls through his business. A man's life is no longer measured by the things that he possesses, but by the sweetness of his spirit.

## Considering the Group's Real Needs

Human resource development is as easy as letting your guard down and finding four friends into whom you can impart Spirit life on a regular basis. Pastor Dowdy has found the key. She is the CEO of a corporation with millions of dollars in land and facility value. Locally, 4,000 people come out to hear her speak each week, yet her joy is in finding a group of peers, CEOs who understand the potential of the resources God has given them and are seeking Him for direction in their use for the kingdom. She has found a new "Life Circle."

Naomi has been accepted in the midst of a group. For the past 20 years of her life she has been the voice over a group. Because she is a woman in a profession historically thought to be male, she has not been in a group. Now, because of the gifts of God in her life and her incredible success, she is recognized by those with responsibility to be a marvelous person. She sees from experience rather than theory and her paradigm is the success of those she leads.

**She has struck upon the second "Life Circle" Principle; she influences this group from within.** She had to first redefine the group in her thinking. Secular employment is not "dirty" or lesser to those who live from the giving of others. It takes no less faith to establish a business today than it does to plant a church or reach a people group. Unfortunately, many of our Christian leaders have never had to face the payroll, make purchasing decisions, procure technical support or compete. In fact, "sheep stealing" and competition are often frowned upon in church circles, but marketing strategies and customer service are watchwords among Christian businessmen. It takes faith to produce and give, anyone can receive.

Naomi leads more by listening to these groups' ideas than by supplying ideas to them. She is able to hear what they have to say. It speaks to her because in her position in ministry, she is faced with many of the same decisions they face. They receive from her a Biblical base of faith for the endeavor in which they are launching. When the waters become tempestuous, they will call her to pray because she is no less a Pastor in their midst than she is in the pulpit, she has gained their respect.

My friend with the resource crisis did not have the respect of the eldership. When the pressure came in finance and development, he turned to a seeming solution, debt. He was willing to sell their future for his success. This destroyed their confidence in him. Any one of them could, with the value of their homes or businesses turn to a lender in time of need, but they have to turn from that temptation realizing that the borrower becomes the servant of the lender. They face it every day; he faced it once in years of ministry. They stood; he could not because he could not tap the strength of their experience.

In an age when *Promise Keepers* and *Focus on the Family* have brought issues of fidelity to the fore, Pastor Greene could not stand. Why? He was unable to step down from the throne and become a part of a "Life Circle." Trained to lead from the front or above, he could not make the transition even though leaders all around the globe were sharing the joys of this transition. The forces that prevented him are those who keep 80 percent of believers outside the doors of local churches. He needed his brothers, but could not come down to lead from within.

The agreed upon solution was to be a part of a small peer group. This will work only if those who come to his side are free themselves from pyramid thinking. If they are in fact able to counsel in the midst rather than over, then he will be restored. If not, their meetings will be formal and the elders will not be included and a bad situation will become worse when, after the year is over, he begins a new career in sales.

Pastor Naomi's group has real needs. They will have to be successful for their effort is in the marketplace where they will be seen and known by all. The others in the marketplace are watching every move. Talk about pressure. They have to think of every word spoken, every promise made, every product sold as a witness to the community. In the city in which they are located, there are four very strong religious groups and Christianity is a minority group. There are forces on every side set to destroy them should they stray. Unlike Pastor Greene's magistrate friend giving subtle warnings, these competitors will love to see them fall.

Because the purpose of the group is to manifest the love of Jesus in helping children meet the challenges of their country; they will need spiritual fuel to succeed. Naomi will give that, as her conversation is the word of God and the Holy Spirit flows from her. She brings perpetual relationship into the group. That level of Christian maturity that does not think of its own success but is mindful that the success of others will flow from her into the other members of the group. It will flow from within a fruit-bearing branch. As they abide in this relationship, they will prosper. She may never make a financial decision in their midst, but she will pray with understanding, affection, and mercy for them. She will cause them to move toward goals of beneficial permanence that will meet their real need.

**When these three principals are manifest — Leadership from Within, Relationships of the Spirit, and Considering the Group's Real Needs — success is certain to follow.** How did 12 men led by a Galilean fisherman turn the world upside down? Their true leader was with them, dwelt in them, never left them, did not use them, equipped them, and moved them to goals of beneficial permanence that met their real need. That was the result of a relationship with Him.

*Chapter 8*

# Material Resource Development

ONE of the great responsibilities which every leader faces is the ability to influence people to give their time, talent, and treasure. All money is in the hands of people. There is no pot of gold at the end of the rainbow, or money tree in the midst of the green grass on the other side of the fence. God has given man the ability to gain wealth. But the balance is, "A man's life worth is not measured by the things which he possesses." It is better to give than receive.

As a result of her acceptance in the midst of a group of very successful businessmen, Naomi will now have to live out a great test. She cannot love them for what they have, or what they can do for her. She must serve them from within the "Life Circle" with no motive to personal gain.

**To build a BRIDGE from where you are to where you want to be, you must have a material resource development plan.** Smart planning counts the cost and develops the resource before digging the hole to build the bridge. More than 70 percent of mission projects never complete the goals for which they were launched. They are very strong in prayer, they recruit wonderful people, they make lofty promises, but they do not have the money. This chapter will suggest several ways in which your "Life Circle" can become a material resource generator to enable people to move toward goals of beneficial permanence which meet the group's real needs.

## Giving in Love

Mother was a part of a "Life Circle." They met at our home from 1955 through 1970. Each Thursday, the ladies would appear. Their meeting was straightforward. We had a piano so they sang a few hymns. Someone taught a Bible study and they prayed. These were three essential disciplines of the group. Then they turned to the real reason for their existence. They loved to meet the needs of younger families in the community.

Some of these women were the wives of prominent businessmen and doctors. Others, like my mother, had to budget very tightly to make ends meet. These were the days of single-income families and the ladies were often restricted in what they could do according to the rules of finance imposed by their husbands. There was no distinction made by amount; they loved each other too much to embarrass a sister. They all participated to some degree.

Over the several years that I was aware of their practice, I began to notice the children of their target families. Many of my peers wore the benefit of their sewing, patching, button repairing, glasses buying, and holiday giving. Our church filled with families who did not fit the 60s Presbyterian small-town mould. We sons and daughters of the "Circle" found in some way the grace to never embarrass these "new" people by admiring our clothes, which they now wore, or mentioning their names to one another. Evidently the prayer part of the "Circle Meeting" worked.

In 1991 my wife, my two sons, and I returned to our hometown from Singapore. We had exhausted our finances in transition but, through the generous help of people in the town, were able to relocate and begin regrouping. A lady called and asked if my wife could join her for lunch one day that week. She had been a dear friend of my mother's, though one of the "new folk." My wife went with her and returned with two closets full of new custom clothing of which the lady had become aware. An old friend of my mother provided the hand money for our home. The local Catholic Church made it possible for us to move in without delay. My office was donated by an Assembly of God Church which had moved into the building left behind by the Presbyterians and this morning as I write this chapter I am in a room in which I received Sunday School instruction as a child, in fact, from my father.

**No gift, given in love, will ever be forgotten in the mind of Christ.** The first principle to a successful material resource development plan is to give. You say you have a need? Give something to someone who has no way to repay you. Give it without their knowing. Give, as these ladies did and God will honor your gift for generations. Give simply, not of constraint, but cheerfully, being mindful that all you have was given to you in Christ.

Your "Life Circle" has been brought together for a purpose. You may be reaching the single-parent moms in your area or the fatherless children. Wonderful, but you will need money to be effective. Not a lot of money perhaps, but some resource. Two things you do not need at this point are a building and high administrative costs. You already have financial commitments to your church, I trust 10 percent, so you must generate funds beyond that. How will the Lord, the "Source" in all "Resource" provide?

**What do you have in your hand?** That is the first question. Rather than thinking about what you could do with what you might receive, think about what you will do with what you do have. One of our friends has a gift for collecting used stuff. Over the years, she has been offered everything from antique furniture to a truckload of Oriental screens and vases that did not sell. Her family loves to hear of mom's latest find.

Each spring she arranges a yard sale and away goes all the stuff. She lets the people buying know to which project the money will be donated and she raises thousands of dollars for needs around the world. She has learned that one man's junk is another's treasure and she finances ministry with the exchange. Does it take time and effort? Sure it does, but she gives what she has to make an impact on her world.

Her son is a missionary in Asia. He has adopted Mom's ability to turn trash into gold and is a legend among his peers for being able to see value in the things others discard. His holiday gift selection is legendary as, for a few dollars, he purchases things which his family will really use daily. Instead of the usual tourist junk, he buys shower slippers and tea so that each morning the family will use the gifts and be mindful of him.

To get to my office I pass through a Sunday School room. At the corner is a bookstand and on the stand is a jar beautifully decorated with the faces of children from the nations. I look for it each time I am in the States. As I pass it, I reach into my pocket and take whatever coins are there. I put them in the jar. Over the course of the year I have no idea how much money goes into that jar, but that class is always first in the monthly mission offering and we as a ministry are now involved in children's work in several Asian countries. Is it coincidence? No. It is a divine principle. Get into the habit of using what is in your hands, or near by, to convert to cash to provide for your vision. The idea that money is going to mystically fall from the sky is an error. But the Lord will bring it to you in the hands of people. Just get yourself into the simplicity of giving and you will see it happen.

## Tithing

**The second principle in successful material resource development is the Tithe.** Let's be clear — tithing is neither Old Testament, nor Law. The first place we see the tithe is between Abraham and Melchizedek, long before Moses climbed the mountain. It was a family thing. Lot had gotten himself in trouble so Abraham went out and delivered him. On the way home they passed the City of Peace and the King came out to greet them. Abraham, from the depth of his heart, gave 10 percent to the King of Peace. There was no law involved. It was in the heart of Abraham.

Since we are Christ's and Abraham's seed according to Paul's letter to the Galatians, then it should be in our heart to do the same. Have you ever asked, "What can I give to Jesus to thank Him for being the King of Peace in my life?" Try the same gift as your father Abraham. Why do the descendents of father Abraham have such acumen for wealth? Perhaps because they were sanctified by the action of their father.

The tithe sanctifies the remaining 90 percent. Many years ago I was called to the bedside of a dying man. Cancer had blown his abdomen up like a big balloon while laying waste to the rest of his body. He rededicated his life to the Lord and, as a part of it, decided that he needed to be a member of a body of believers. He had asked me to which church he should belong, so I mentioned one Pastor as being a person of relationship. This man's family was going to need such a person because he was soon to leave this life.

I didn't say anything to the Pastor about our conversation as the man was very wealthy and I didn't want any question of motives to enter into what was a pure relationship. To my surprise, several days later I received a call from the Pastor. The man had gotten out of bed, called his driver and gone to the offices of the church. He had prayed with the Pastor and, in leaving, handed him a check. I did not know, but the church needed $350,000 for the balloon payment on their land. The man had felt the need to tithe on his earnings so far that year. You guessed it, on earnings of 3.5 million, he gave the amount of the payment. You could say that the Pastor was in shock, but I was overwhelmed. I honestly had no idea of the man's wealth or the Church's need. I know that the man giving the tithe had no idea of the magnitude of his gift.

The man passed away shortly thereafter and his wife continued in the church. There came a financial crisis to Asia, but their holdings were not affected. The tithe sanctifies the remaining 90 percent.

If your "Life Circle" can embrace this concept, then I have a radical suggestion for you. Why not bring ten percent into the storehouse of the local church as "The Tithe" and another ten percent to the "Life Circle" as "a tithe"? Why not try to get to the point at which you can live on 80 percent of your income. That will cause some changes, but if a group of ladies in the 60s can do it, then so can you. If you have ten incomes in your "Life Circle" and they tithe, you can support one of the ten for a month. That one could go for a mission trip, volunteer with Habitat for Humanity, spend a month supporting the families of children undergoing surgery, go to a conference to get more equipping for the group, volunteer at your church office, go sit with an aged relative. You see, the point is all money represents time at some exchange rate.

Money will buy time and time will provide money. When a person makes a cash gift to a ministry it represents some amount of time that they have spent working at some wage. A friend of ours is a dentist. His specialty is root canals. I cannot imagine sticking my face in someone's mouth all day and digging out decay. He doesn't particularly enjoy it either, but it is what he does and he is very good at it.

He once explained to me while donating a sum to this ministry that in his mind he was doing that root canal to reach children in Asia with the Gospel. Suddenly it gave additional value to his work. He could see divine purpose coming from the stench of decay. He was turning hours of his life into cash, which would buy eternity for others. What a refreshing way to look at the way we work.

As Pastor Naomi spends time with these emerging business leaders, she is going to realize the value of one hour of these people's time. At their incomes, one hour can mean thousands of dollars. But, that hour of life is no more valuable to them than an hour is to you because our value is not measured in what we have, but in what we give.

Pastor Greene was seen as the income generator of the church. The weight of the buildings and the outreaches fell squarely on his shoulders. Eventually the weight became too great and it pressed him into a sinful state. It was to his amazement that the businessmen who served as elders came to cover him. Would he have done the same for them?

Remember my friend with the cash crisis? He went the same way. Instead of developing a relational line across which resource could flow, he became the key man in a pyramid, which eventually caused the destruction of his ministry. What was his misperception in resource that caused the problem? He failed to see the value of the multitude of counsel. He became frustrated at the speed with which the board moved in an important financial decision. Time became a factor, as he wanted to accomplish his purpose quickly.

Time, when multiplied with Selfishness and depleted Trust, led to the point where there was insufficient trust to continue.

## Trusting in God's Provisions

Material resource development can be the boon or bust of any ministry. When faced with a financial need, Hudson Taylor went to prayer. As recorded in the book *The Key to the Missionary Problem* by Andrew Murray, Taylor needed $50,000 to be able to keep China Inland Mission afloat and debt free.

Taylor had limited time to act, so he went to prayer. "Lord," he said. "I need $50,000 to operate. I haven't time to contact many people, or to write many "Thank You" cards. "Could you please touch the hearts of five people to give ten thousand dollars each? Thank you."

He went about his normal routine, not altering schedule or demeanor and, within the week, he had received the five gifts. How is this possible?

**First, he went to the true source of all provision.** He prayed. He called on the Lord.

**Second, he remained himself.** In today's world of marketing the church we have to create a crisis for people to respond. "If we do not hear from you in the next ten days, this ministry will end." Perhaps it might as well end then. Taylor did not create emotional appeal to manipulate people into giving. There were no midnight voices directing him to contact five people, in fact he slept quite well, resting in the promise of the Lord.

**Third, he kept the need between himself and his God.** This is the greatest challenge for all of us who head ministries. The burden of funding comes to us unless we are surrounded by a "Life Circle." Taylor moved in such circles. He was not a fan of the Mission Society, but rather spent his time with those who, in obedience to the call of God, lived a simple lifestyle of service.

**Fourth, as the money came, he did not try to "leverage" it into a great cash flow.** He stuck to the request he had made of God and gave Him the glory for the provision.

Pastor Dowdy moves among millions of dollars. She lives and works in a city where people spend US$125,000 on a BMW. They think little of club membership fees of US$20,000. In the face of all this, she does not live an excessive lifestyle. Her staff represents Jesus with humility, which they have inherited through her example. In the face of tremendous financial need, they have learned to call upon the Lord and have seen His abundant supply. They are a "Life Circle" for each other.

## How Are You Faring?

Let's look at your plan. First, do you tithe? That is going to be an issue. See it as worship as Abraham did (Genesis 14: 20).

Second, are you sowing seed into the area you want to reach? Remember money is time at an exchange rate. You do not have money yet? Give your time and watch the Lord's exchange rate.

Third, are you trying to carry the need yourself or are you involving a "Life Circle" through inspiration, information, and invitation? One of the tests you will go through before the Lord opens the floodgates will be your ability to share the ministry with others. If it is "your ministry," you will carry it. If you surrender it to the "Life Circle," the Lord will carry it. It is your choice. That ark would have been really heavy for one person.

Fourth, what do you have in your hand? Is there something the Lord has already provided to meet the need? How much of what you have do you need? Are there a few bags on the camel that are preventing it from fitting through the door?

Pastor Greene is learning a very valuable lesson. Even though he has "fallen" he is going to be sustained through the continued giving of the men who have come to his side. In being humbled, he has been placed in a position for future ministry. His pyramid has become a "Life Circle" through God's grace. The same provision that enabled him to become a valuable asset in the community will now carry him to an even greater position, humility.

It is too late for the other church. There is no "Life Circle" forming there. The elders are tired after years of tension. They no longer have the love to press on regardless and are leaving the partial Circle they once had. The final stone has sealed the tomb.

Naomi wins and the Kingdom of God wins through her. She has made the transition, departing from the pinnacle of pyramid perfection for the service role the Lord intended. Her "Life Circle" will grow in resource and result as together they seek the Lord for His way and timing. Their city will continue to be affected for Christ. Because they have chosen the way of service, others will be drawn to them.

Her gift for inspiration will become entwined with their abilities to inform and invite, forming a three-strand chord over which all resource will flow. Their "Circle" will reach a nation where my Mother's reached a neighborhood.

You will see spiritual, human, and material resource flow to your vision as you follow the example of Christ in forming a "Life Circle." As you inspire, inform, and invite others to join you, there will be a multiplication that will eventually reach the globe. Be confident. Start today. Do not seek to be over them in ministry, seek to serve from the center and allow the circle to form. God will bless it and you.

*Chapter 9*

# Intercession

"BEFORE this morning's session I met with the committee. I asked a very simple question of them and was surprised at their answer. I must say I applaud their honesty, and when I asked them if I could share the result with you all, they agreed." Pastor Edmond had my attention. We were starting the third morning of a Pastor's Prayer Summit. There were 137 churches represented by over 650 delegates. Called together to begin the New Year with prayer, I was amazed at what came next.

"The question I asked was this, 'How many of you have a habit of taking one hour each day for prayer?' " he shared before all of us. "And the answer was, 'None of us.' Not one of those who you all trust as leaders prays for an hour each day."

## The Importance of Prayer

**The "I" in our acrostic stands for Intercession.** Webster's Dictionary defines Intercession as "Prayer". That is all it says. For the first definition of the word, it just says "Prayer". Is it any wonder that Pastor Greene was found in the darkness of sin and despair?

Prayer ministers to the spirit and the flesh of man. The sensations of intimacy in the prayer closet can touch our deepest desires. Men approach prayer as work. Women as adoration. It is difficult for a man to think of intimacy with another man. Women, on the other hand, find in time with Christ an understanding that no human can produce. As a result we find the prayer life of the church delegated often to the women. Men have more work-oriented things to do.

**The success of our "Life Circle" reaching its new destination will be in proportion to the amount of time we spend in prayer.** As the Director of the Operation Unreached ministry of the Association of International Missions Services, I was required to familiarize myself with the 7,000 plans to reach the remaining unreached peoples of the world by the year 2000. In looking through them I realized two common denominators.

First, someone had to go. It is the same message Jesus gave the disciples — to get from A to B someone has to leave all and go forth with the Good News.

The second was no easier than the first; God had to do the lion's share of the work. God had to, by His Spirit, draw all men to Christ. That would happen, came the conclusion of the 7,000 strategy writers, only in response to believing prayer. Hence, the prayer movement of the '90s. Thus was birthed the Prayer Guides, Prayer Maps, Prayer Centers, 10-40 Window Prayer and many ministries called, not unlike our own, to reach the nations through prayer.

How then could the leaders of a revival in the Third World not have a prayer life that included one hour of daily devotion to the presence of God? Easy to understand, prayer is not a movement, it is a personal decision.

## The Wrong Way to Do It

Prayer requires us to leave the affairs of this life and "enter in" to the presence of God. It is this "entering in" process which keeps pyramid dwellers from participating. As a rule, they delegate prayer to those who have the time to spend in quiet, listening to the voice of Christ. These small groups of intercessors are kept in check with just enough information to pray, but not influence the direction of the congregation. The group is often composed of middle-aged to elderly women who generally love, are kind, and are deeply devoted to the Pastor. They meet at the church and are "under the covering" of the Pastor. They become his insulators to the events and lives of those "under him" in the ministry.

By delegating the prayer ministry, the leader avoids conflict with four forces which withstand the ministries' growth and souls being reached. The church does not enter into the community as a "Life Circle," giving life to others. Instead, from its location in the safe suburbs, it plans occasional outreaches to the population center. This isolation communicates to the community as disinterest.

Pastor Jon's financial problem came as the result of this prayer delegation. Instead of prioritizing his elders' meetings for "Life Circle" development with a prayer emphasis, he allowed them to become business meetings. He felt that the prayer ministry of the church could be better handled by the more spiritual intercessors, and that the material decisions would be best served by the eldership.

They wanted to reach the community through material or humanitarian ministry. The Pastor wanted to have the people come to the church to receive their blessing. He wanted to build an addition to the existing church/school. The elders suggested that each of them use his garage as an extension of the church and reach the community. Because Spiritual Resource Development had been delegated, they could not reach unity on Material Resources.

He went privately to a bank to obtain the funds, but the banker, who also served one of the elders, had asked casually why the church was expanding their facility. What was done in private was then revealed publicly. The elders took great exception to his independent act. He took great exception to their private meeting with the bank. Trust ended.

## The Right Way to Do It

**Webster's second definition for Intercession is "To intervene between two with a view of reconciling their differences."** This is what your "Life Circle" is going to have to do to move toward goals of beneficial permanence that will meet the group's real needs. It is this that Pastor Naomi is doing in the business world. She is leaving the building to enter in to the community.

We have been given the ministry of reconciliation. This is the reason the Lord has caused your "Life Circle" to form. This is the dynamic force of purpose that will cause Bob and Lesley and their Pastor to reconcile and work together. This is the weight which the body of Christ is designed to carry and for which it is being strengthened through that which every joint supplies. God wants to be reconciled with man. That is His purpose, which governs the events of all human kind.

There are four forces, which resist this purpose. Your entering into the area to which your "Life Circle" has been called will bring the manifestation of these resistant forces. They are the key issues which work to prevent reconciliation. As we look at them you will be able to see why Pastor Greene found himself in a patrol car and what will have to be overcome in his life for him to be restored. You will understand why "Relational Teams" so often give way to pyramids, why resource stops flowing, and how to overcome.

Whether you are reaching out to single Moms, little children, government leaders, or the business community, you will encounter these forces. I have found them in Kings' Palaces and paupers' hovels. They are common to the earth and, in understanding them, you can influence your "Life Circle" to remove them from the lives of those you seek to save.

## Resistant Forces

### Who Was Here First?

The first of these forces is the thought, **"Who was here first?"** or "Whose territory is this anyway?" You may recall the musical *West Side Story* which first brought to the screen the idea of gang dominance in New York and touched on the Shakespearean theme of cross-cultural romance. Today the world is being divided into ethnic states. Thousands have died and many more will as the question of territory pits neighbor against neighbor. The brutality is now called "ethnic cleansing." It used to be "genocide." What ever it is called, it is the manifestation of what the Biblical Greek calls "Archas." The issue is who was here first and whose is the world.

The Gospel of John begins, "In the beginning was the word and the word was with God and the word was God." The word "beginning" is the Greek "archas." It carries a sense of preeminence. If God was here first, then the earth is His, and we are creations in it. If God is the product of the fears of man, as some would assert, then man was first, the world is his, and God is a creation in it.

As your "Life Circle" tries to get from A to B you will wrestle with this thought, "Are you creating a ministry and asking God to bless it, or has God created the ministry and you are walking in it?" Pastor Greene saw the ministry as his. "People drive 50 miles to hear him preach. After all the church was called by his name." What had happened?

Without the humility of prayer, any one of us can become self-sufficient and fall to the cunning craftiness of this force. We can begin to think we are good. We can even begin to feel good about serving the Lord. We lose sight of our total need for Him. Without Christ we are bankrupt in the earth. He was here first. He will be here after we are gone. He sustains all things. It is not the work we are doing for Him; we are the work He is doing in the earth.

To whom does territory belong? I used to book all of my travel with the same airlines. As a result of the accumulated miles I received certain favorable treatment. I saw it as good business. Then the airline's regional sales manager explained to me over lunch that they were going to eliminate Christian consolidators. These are those wonderful folk who live to provide missionaries with a low airfare. Their joy is found in the total amount of money they have saved the body of Christ each year and the number of nations they have reached through serving others. Upon receiving this information, I changed airlines.

About a year later, I had occasion to fly on the former carrier, as the schedule demanded it. I was relegated to a lower class of service until my name was registered in the computer and then the manager of the ground staff apologized to me and asked if I would like to join the preferred line. I glanced at those about me. Their faces told a tale of aggravation and frustration. Why should I be elevated? Wasn't this territory good enough for me?

I declined. I could hear the sound of my own voice in the concern of those in the preferred line. I was ashamed as I heard the arrogance that had been mine. I saw the frustration of the employees as they attempted to serve within the limits of the seating.

The issues were all preference and priority. "Do you belong in this space?" Now my world of travel might seem strange to you, but what about the church. What about your office or business environment? What about your family structure, personal finance, sibling rivalries, rooms and space allocation, budget priorities?How are all these being assigned? Do you pray about it all, or do you decide on, "Who was here first?"

**New people are joined to churches because old people do not have all the answers.**

Having lived through three revivals, I have seen firsthand how the old folk relate to the new folk. Who sits where in church is just one thing. What about the charismatic couple who find their way into an evangelical church? Or, better still, the Catholics who tenderly try to see if they could have a place in the new Community Church.

I can recall when the "Real Gifts" of the Holy Spirit were the exclusive possession of the Assemblies and when the True Jesus Church began. This issue of "Who was here first" has such a hold on us that the greatest resistance to the Unreached Peoples movement became the existing mission ministry of the church. How often has your great idea been rejected by, "We have seen that before?"

We work in South-East Asia where to be Thai is to be Buddhist and to be Nepali is to be Hindu. We still have the wars of Hindu India and Moslem Pakistan and, just now, thousands are dying in Indonesia over land which was transferred by the central government to new owners from other people groups. The scenes of Kosovo and Croatia will be with us for generations. Upon this canvas of "Archas" is the picture of Christ painted in the blood of the martyrs.

He who was from the beginning reaches out through you to embrace a people who have not known Him. You will be His hands of compassion reaching to them, meeting their needs through His provision. As you enter in to their environment, you will wrestle with the "Archas." Have confidence. Through Prayer you will be victorious in your goal to move the group. As you take the servant role in their midst and distribute spiritual, human, and material resource, you will drive off this force. But, there is another you will face. It is a little more difficult.

### The Right to Speak

"Do you have a permit to meet here?" the young man asked a common question in his native China. "Do you have permission to speak publicly?"

"Yes," I answered, as I led the mime troupe along the crowded street.

"Did you get it here?" he asked, matching his stride to mine.

"No, higher," I responded and, satisfied, he turned and left us to ourselves.

**The "Right to Speak" is going to be another force with which you wrestle as you enter in to that group or place of ministry to which the Lord is calling your "Life Circle."** You are headed for the inquisition of your life. Pray a lot and, when people you love oppose you, love them in return.

The Greek word is "Exousias." It is the word used in the famous commission of Christ, "All authority has been given to me in heaven and on earth. Go therefore and make disciples of all the nations, ..." (Matthew 28: 18, 19a). It is the word "Authority." Simply, it means, "The right to speak." Jesus has the right to tell us what to do because He made us in the first place and redeemed us in the second. We are His possession by purchase so He can tell us what to do.

"Fair enough" you say, "but not so fast." If that is true, then why do we not go into all the earth and make disciples of every nation? Is it because we are under the influence of this force? Yes, it is true. Each of us wrestles with the right to express ourselves. Here in America this right has been perverted into forms of expression which are terrible. However, the right is precious.

Pastor Greene is going to hear from a group of people to whom he has never before given the "right to speak." They are, in his moment of personal shame, covering him before the press, but they will speak into his life. His wife has not yet spoken, but she will now have the right. The community to which he has preached purity will now have the right to respond. Some will be very cutting. Hopefully their cutting comments will be enough to cleanse the cause of his sin.

Naomi has opened the right to speak to the business community in her congregation. She will soon know how relevant her teaching has been to their lives because she will hear her own voice in what they have to say. If she listens closely, she will be able to evaluate her ability to communicate the Word of the Lord to them.

When Bob's pastor would not take his call, he sent the message of the "exousias," "Who are you to interrupt me?" It really hurt. Bob got the message. There was not sufficient relationship for communication to be bilateral.

When your "Life Circle" meets, everyone has the opportunity to express that which they feel the Lord has given to them. As we worked on the Bible Training Center, I was very impressed with the things the others in the group heard from the Lord. One of the brothers was an educator, his understanding of learning and retention was particularly wonderful and we naturally looked to him as a guide in our discussions. I observed, however, that his education was not his greatest strength. He was a master at drawing out the quiet members of the group and some of the best input we received came from those whose education was life itself. As a servant to the circle, I rejoiced to see those who knew how to defer to those who were learning so that the whole would benefit.

When you see a struggle begin over the right to speak, begin to pray. This is a force. However, it will be overcome in three steps.

**First, pray that the voice of God will resonate in the hearts of those involved.** When God speaks, the earth stands still. He has the right to speak, for He spoke and the worlds were formed. He can make Himself heard.

**Second, check yourself out. Are you a better listener than a speaker?** Are you demonstrating to the group the gentle way of listening to others, or have you given over to the siren song of external authority and are you trying to control through inflexion or word choice?

This is an issue. The temptation to power struggles is always there. We wrestle with it. Jesus, who knew all things, brought every concept of the Kingdom of God down to the simplest terms so that we can understand. We, as members of "Life Circles" must be ready to do the same. If you, by virtue of education or understanding have information the others do not, then find a simple and sincere way to communicate it.

**Third, encourage others to share.** Pray that the Lord will give you the grace to think of them more highly than you think of yourself. Comparison is the basis for condemnation. You will either be lifted up in self-defense or put down in self-denial if you wallow in the bog of comparison. The Bible tells us not to be like those who compare themselves with themselves, because they are not wise (2 Corinthians 10:12b). "If any of you lacks wisdom, let him ask of God, who gives to all liberally and without reproach..." (James 1:5).

Imagine what could have happened if my friend, Pastor Jon, had heard the suggestion of the board and allowed resource to flow from the central church into their homes to be distributed into the communities. He could have, from the central position of server, multiplied the church five times over in about six months. He could have reached those who do not feel they have a place in the church. Those who cannot make it to church because they do not have transportation, or even those who do not feel that they fit in to church because they do not know what to say when they get there.

**Actually, every place we find the words "Powers" and "Principalities" in the New Testament, they are the two Greek words, "archas" and "exousias".** You have just read a great key to success in reaching goals of beneficial permanence that will meet the group's real needs. When you give place to your brother and allow him to speak, you prevent the types of problems that Bob and Lesley have faced and the torturous pain that Lisa Greene is facing.

## The World's Way

**The third force is also obvious if we just stop to think and pray.** It is the rule of the world's system. John Bunyan wrote of Worldly Ways and his attempts to stop Pilgrim along his journey. The Greek word here is "Kosmokratos." Debt or no debt, that is the question. The borrower is the servant of the lender in every case. As you developed your material resource plan, you realized that this would take time. Your local lending institution is not the money tree for which you have been looking.

Jesus is the source of all resource. His name is Provider. While serving in a large church, I was shocked one staff meeting to hear that there was not enough money for ministry budgets so do not even turn them in. I was responsible for ministry in several areas, not the least of which was Children's ministry. Visions of Sunday without apple juice and peanut butter flashed before my eyes. Thoughts of six hundred kids all charging me for their snack and handicraft caused me to toss and turn through sleepless nights. What if they all turned on me at once and rushed the platform? I could be crushed against the back wall like a dead fly. What to do? What to do?

I called our Children's Core or "Life Circle" together and painted the picture for them. We prayed together. We implored the Lord with all zeal as our lives depended on His provision. He answered. He provided. I cannot tell you exactly where it all came from, but we never lacked for apple juice and peanut butter and jelly. The Lord was faithful. We ministered for that year to those kids for less than $70.00. That's right, less than seventy US Dollars.

When the world's pressure surrounds you, do not turn as my Pastor friend did to the banks. Get with Jesus. Give out of your need. Be sure you are all tithing as a principle. Invest what you can find in the ministry. Pour out what you have to be ready to receive what you need. The Tithe is the way that the devourer of interest is kept away from your finances. Sanctify the total by giving and make it very generous.

## Spirits of Malice

That's three down and you are almost home. If your "Life Circle" can recognize the fourth force and take up a prayer position against it, you are going to impact thousands for Christ. The fourth force is terrible. It comes in clusters. It moves in the most serpentine ways for, indeed, it is translated, "Spirits of Malice." I hate to have to give them page space, but I would not be giving you the whole picture if I failed to mention these.

What entered the mind of Hitler to hate Jews? Why do the Moslems in Indonesia play with the word "Jihad" or holy war? Why does a young man enter a secondary school and start killing people? What would be so devilish that a brother would rape his sister or a father his own daughter? Spirits of Malice have invaded our lives. They occupy the front pages of most papers and "inspire" much of what is in the media today.

**The Greek word for these is "Pneumatika taes Ponerias."** There is little doubt that these agents of Satan are the angels who, with him, were cast down from the heavens and have become demon forces. They stand in the way of the completion of your BRIDGE. They oppose everything for which you stand. They knew just where Pastor Greene would feel good. They know everything about you and exactly where to lurk so that in a weak moment you can be compromised.

Even the President of the United States was not safe. Neither were students and faculty in secondary schools, nor women and children in Bosnia. Little children playing in villages in Laos were burned horribly as American planes dropped "napalm" and "agent orange" on them. I have been asked often, "Why?" As I travel there in our "Life Circle" for South-East Asia, I am hard pressed to answer.

"If these are the actions of a Christian nation, why should we receive Christ?" is to me a reasonable question even though I am a veteran and very happy to be an American. The forces that cause us to hurt one another in a malicious or preplanned way can be defeated only through prayer.

As we intercede for the nations or the group which we desire to reach, the Lord does several things in our hearts.

**First, He changes us.** He releases in us that grace which we have received in Christ so that we can love beyond human ability. To the theologian, we go from Phileo or brotherly love into the dimension of Agape or perfect, Christ-given love. This is a must if through this comfort which we have received we are going to comfort others.

**Second, He cleanses the lenses of our understanding.** That is to say that Jesus causes us to see people as He does. We forgive and are forgiven. I have often asked Vietnamese and Lao people if they have ever had the opportunity to say to an American all they would like to about the things they have suffered as targets of our bombs and agents. Most of these folk were very young during the war and their only remembrance is the insignia on the wings or body of aircraft and the resolve to some day get even.

I wait and even encourage them, telling the translator to not spare my feelings but to tell me exactly what they say. Some have been very gentle in the Asian way, but others have taken advantage of my openness and really blasted away. At the end of it all, I explain to them that I always had a hard time forgiving, but after I received Christ I was able. And in most cases I have been able to ask that person to forgive me and America and they have. In most, though not all cases, I have been able to pray with them to receive Christ and we have continued some form of relationship.

Love cannot be defeated. Where these forces have operated, prayer must bring the release of love that will set the captive free. Through prayer you can defeat this force.

In our book *Every Place Your Foot Shall Tread*, I encourage the reader to go to five places in their town to pray. I explain in detail that the glory of God will flow through you. For our purpose here of bridging the gap from where you are to where you want to be, let me just give you the five places — Centers of Government, Centers of Commerce, Centers of Communication, Centers of Education, and Centers of Spiritual Activity. Go to these places regularly and pray that the Lord will rise up through you and your "Life Circle" and will defeat these four forces.

## He Is Able

He will do it. He will move through the group. You will testify as the disciples did in Luke 10, "Lord, even the devils are subject unto us in your Name."

You must get out into your target area and ask the Lord to clarify the vision your "Life Circle" has received in prayer and give you the specifics you need to be effective in achieving your goals. He will give you the understanding necessary to reach that community or segment of the community for which you are asking.

"Who was here first," "The right to speak," "The World's Way," and "Spirits of Malice" — all four will bow down to the Name of Jesus.

Pastor Edmond stopped the summit and called us to prayer. We all prayed for an hour. He did not allow speaking or sharing any vision for the hour. There we were, more than five hundred leaders on our faces before the Lord; the elders of the body of Christ in that city humbled before the Lord. I have no doubt that when we gather next year, he will again ask the question.

*Chapter 10*

# Practical Prayer Power

AS we pray, the Holy Spirit forms a veil over us. I have no idea how He does that. No one could ever explain the warmth and peace of his manifest presence, but it does form a shield of protection. The bold declaration of Colossians 3, "You are hidden with Christ in God..." becomes a reality.

As Jesus makes intercession for us, we bask in the radiance of His presence. He becomes a constant force in our lives. So it is when we begin to pray for others. We invoke the name of Jesus and His powerful presence produces peace in their lives.

The Ark of the Covenant is a picture of this abiding presence. Borne by the priests, it was the residence of the Glory of God. Wherever it was carried, the glory came to bear on the forces that were resisting the purpose of God. As we pray and enter in to those areas we wish to reach, we bring upon the altar of our intercession the anointing which will cause the forces to bow.

As Pastor Greene wept before the board of his church that dismal morning, they too felt the cold conviction of the Holy Spirit. What had been their failure? What had allowed the enemy to drive so deeply into their midst? They realized they had not been faithful in prayer. In their early morning meeting they had confessed that many weeks had gone by in which not one of them had lifted their Pastor in prayer.

As they had reacted to the pain of the Pastor's fall, they had asked each other about the power of prayer. They would certainly need it now. They would be facing a very hostile press. As they prepared, they questioned who would speak. What did they have the right to say? Who was first among them? They had almost followed the way of the pyramid in their decision and then they had stopped to pray. They asked the Lord for His forgiveness of their sin of omission.

Gene, the quiet dentist, had felt the leading to pray in this way. Humble, always ready to look at his own sin first, he had suggested that they had dropped the ark. They had failed to carry their Pastor in intercession. They had taken for granted the blessing of the Lord and had been lulled to sleep by the apprarent prosperity of the work. His prayer of repentence had been from a broken heart. He had really believed that Pastor Greene was above sin. He had confided in the man his own desires for women and they had prayed together for his marriage. He had never once thought that the Pastor might be having those same desires. Now, in the cold clear light of dawn he saw his sin. He had failed to cover and sustain his brother by lifting him to the top of the pyramid.

Luke had been slower to acknowledge their lack of power. He had a Morning Prayer discipline. He had prayed every day for the Pastor. Their recent gentle conflicts had resulted in his almost leaving the board. His wife had encouraged him to stick with it as he needed to follow up his prayers with expressed love. He hadn't been excited about the prospect and had been recently in the habit of sitting quietly in his office asking the Lord what he should do.

Now, as he prepared to face the press, he realized that the disturbance in his spirit had been the Lord letting him know that something was amiss. How he wished he had been discerning enough to follow through in prayer and draw closer to the man he was now going to protect. Was it too late? What would happen to the church? What was he going to do? Would his reputation and business be affected by what he was about to do?

All these questions assaulted him as he prayed about what to say. He was about to "enter in" between the Pastor and the community with a view of reconciling their differences. How would it all work out? He had no idea, but he was going to find out.

## The Enabling Power of Prayer

Lisa had prayed. She had spent sleepless nights in prayer. Her anguished soul had cried out until there were no more tears. He hadn't come home. He had not called. When he did come in, she could smell perfume, not her own. She knew.

For months she had taken her place on the front pew, the professional smile on her face and ice in her heart. She could not take her pain to anyone in the town, for he was Moses leading the people. He prayed for them and they were healed. He taught them from the pulpit and they sat in awe of the revelation he received. He spoke to them of love and purity and they vowed to change their lives. And all the while she knew. Deep in her heart, she knew.

The situation had formed a "Life Circle" in her living room that morning. It was almost complete. Still to join were the two Pastors and the three children. How would they survive their father's fall?

As she looked about the room, she wondered at the depth of the commitments among them. She really didn't know the Board at all. She had tried to reach out to their wives, but always got the feeling that the ladies were trying to get information from her that they could then repeat. Perhaps the distrust was unfounded, but she would soon find out. She was sure they would be involved in the "Restoration" process.

Wow! Imagine that, she and her husband were going to be restored. To what? What mold would be deemed perfect? How would her behavior be analyzed? Who would dig around in her eye to find some speck of dust that had caused the Pastor to wander? Would she get to choose, or like everything else in the church, would she be told what to do?

She could feel the years of hidden resentment rising up in her throat. "Not here" she steeled her nerves. "Do not break down in front of these people. Do not let them see how deeply you are hurt. Hold on to the Lord. He will not forsake you." Her fingers dug deep into his shoulder. Her hands would not shake. Her face would not flush. She had done it for years, she could do it now. The chill of pain formed icy conviction as she prepared to face certain shame.

She had prayed for him every day. In fact, it was her prayer life that had kept her from leaving the first time he lied to her. The worst had been the nights he did not come home. The agony had become a garden of Gethsemane to her. She could not open her heart to a soul. Thank God the children had slept through the nights. Jainine, usually sensitive to her needs had been spared and James slept like a rock.

Megan had been her companion. Was this the reason for her illness? Many would say so once his sin was declared. They would say that the illness caused the sin or the sin caused the illness. They would never know how she had clutched that child to her abandoned bosom and had rocked her the endless hours. As Megan's breath came in tortured gasps, so hers began to come now. She was going to lose it.

Bill saw her need. He came across the room with gentleness that very few realized lived under that builder's gruff scowl. "It is going to be difficult, Lisa, but we will be with you. We will not sacrifice you to the wolves. We are here to surround you and speak life into this day of death." His strong shoulder became the reservoir for her tears as heavily muscled arms shielded her from Lucifer's lash.

"No one is going to be able to hurt you," Gene joined in. "I promise you that the church is going to cover you two with love as you have never known it. We have seen this morning that we have been remiss in not covering you with a shield of prayer and concern. We have taken from you all you have to give and have not given in return. Please forgive us."

His sincerity really broke her. Gene and his wife, Marge, had always been ready to help her with any need. One of the first couples to join with them, their friendship had always been a source of strength to her. Perhaps she could go to Marge. Lisa felt much better. Gene's calm became her shade from the blast of morning misery.

Robert Greene could not respond. He sat on the sofa emotionless. He stared at the curtained window. He had no means of escape. He loathed these men. All their words of love meant little to him. What did they know? How could they understand the pressures of leading the flock? What did they know about him, his wife, and his bed? They knew nothing, but in their usual "do good" mentality they were here.

Love, what was love? He felt no love. The Judge had betrayed him, and what about that police chief. If he told the things that man had confessed he could have the man fired. If these men knew half of what he knew, they would walk out the door. He had not felt love in this home for the past five years. There was no love here, only questions and longing glances. How could they know? Restoration to what? Who would they appoint to look into his eyes and try to find the truth? No Pastor in the community had the ministry track record he did. There was no peer. There were only those who made a good imitation of his style and model. He was the leader and they would soon see it.

The doorbell interrupted his thoughts. "Oh great, here come the two thieves," he thought. "Soon the faithful would migrate to their pews and be comforted with their fine words."

Lisa smoothed her hair and, wiping salty tears from her eyes, answered the call to service. It was, after all, her home.

Her regal bearing almost hid the pain and embarrassment in her heart. She was radiant. She was instantly in control.

"How," Robert wondered, "can she always do that? Where does she get the strength? The tilt of the head, confident but without arrogance. The set of her chin was the prow of a ship ready to handle the next wave. Her shoulders were set in gentle strength. How many times he had seen it.

"How can she do that?" he continued to marvel as the two men received her greeting with cheek busses and came into the room. He could not take his eyes from her.

This was not his wife. This was the daughter of God; broken, embarrassed, humbled, but in control. Her moment of weakness past, she was now more than a conqueror. He marveled as she gracefully came across the room to stand once again behind him.

The two men declined her offer for coffee or tea, and took

their places in the circle. It was like Shakespeare in slow motion. The stage was set, the principles were gathered, the crier had made his cry and now the actors began their dialogue. To Robert Greene it seemed a farce. He could not get into it at all. The old drive to survive took over and he prepared to address the group.

"Robert," he was interrupted before he began. "We have come to stand with you in this hour. Please, do not think of us as "judges" or "thieves". For once we have come to give to you, not to take from you. Please accept our visit as friendship, even though that is a new term for us."

## The Loving Power of Prayer

What was this, a soft word instead of judgment? What was the game, when would they attack? They had to attack. He would have. He would probe deep into it and get enough information to control the situation for years. What were these two doing?

"We want to form a circle around you which will include the two of us, your board, your wife, and whatever professional help we all feel is necessary. Will that be all right with you?"

A circle, what in the world, were they talking about? Was this some kind of a new age trick? Was it their way of saying, "You are not the top man any more?" Who did they think they were? Sure it was a rough day, but this had drastic implications. Soon they would want him to spill all the corrupted content of his lust-laden soul. What would they do with all of that? Wait 'till they heard the things women had done for him, that would end their circle soon.

"Sure," he responded. "Whatever you think is the best thing to do. I will submit to your direction in everything. Thank you for coming."

Lisa could feel her skin crawl. He was recovering control. How many times had he said essentially the same thing to her? "Sure, Lis, whatever you think is good." She had heard it through every situation. He would manipulate them the same way he had her for so many years. Life at the top had taught him many ways to control.

"Bob," Bill looked him in the eye. "Why don't you cut the crap and get down to it. You are finished. Your ministry is done. Your family is going to make very tough decisions and you better be ready for hell itself to camp on your door-step. You think you are the only person that sinned around here? These people can tear you apart with a look. You need this circle more than you could imagine. From your protected pulpit you have never had to feel the ire of the congregation. Buddy, they can and will make you feel like dirt." How he had held the emotion was a mystery to the others. Bill did not tolerate any smug attitudes in his business and had read Robert's from the outset.

"It will be only through believing prayer that this horrible reproach can be washed from the life of the church in our city." Asa was the senior Pastor in the area and an official in his denomination. Pastor Greene had often bristled when he discovered that members of his church had gone to Asa for counseling. Several of those who had suggested that there were problems in his marriage had become members of Asa's First Assembly.

"In fact," he continued, "Let us pray right now."

## The Cleansing Power of Prayer

Nodded agreement brought the simple flow of cleansing as one by one they spoke their sense of weakness to God. Lisa alone did not offer her thoughts publicly in prayer. As they raised their eyes, hers met Bill's. He smiled with a Spirit-borne love. It was pure, it was direct, and it was love.

Pastor George Hall was a tall man. He often sat down to talk with people so that they would not feel intimidated by his physical presence. Folks in the town said that he was once a violent man involved in bar-clearing fights. Actually, he was a gentle giant. He stood to speak to Robert, "Bob, we talked quite a bit on the way over here, Asa and me. We have for several years been a part of a small, private, prayer group. You know some of the men."

"Yes," thought Pastor Greene. "I know them very well."

"Asa and I thought that perhaps you would like to meet with the group as your actions will have an impact on them all. Our only consideration was whether you were ready for the transparency with which we relate. There is no head to the group; we are not called together on the basis of denomination. In fact our denominational leadership is, in most cases, unaware of the meeting.

"We are getting together today at three. It would be a good opportunity for you to confess before your peers before the evening news is out. It would be a good first step for the healing process to begin."

Pastor George stood silent, gazing down at his crestfallen colleague. He did not milk the moment, but with sincere pain in his eyes implored the face before him.

Robert Greene went through changes. He had laughed often about this group. "Losers," he had called them and had taken great strides to avoid their meetings. Their phone numbers were not on his speed dial. Now he was being invited to confess his sins before them. Who were they to sit in judgment over him?

"I think that would be a great idea," her poise was still amazing.

"It will give me the opportunity to sit with the kids and make our 'help Daddy' plan."

"Well, you said you would submit to whatever we directed." Asa added, looking for a bit more enthusiasm in the response. "If you're going to be brought down to live with the rest of us, then you might as well take the tough tumble first."

"Are you sure you would not like to have that coffee," Lisa asked now thrilled that her husband had been backed into a corner from which he could not escape. She even began to hum as she went to the kitchen. "You men work on the public statement and I'll fix you a snack. How many want decaf?"

Luke was very nervous. The morning's valor had passed as he sat and watched his Pastor. Did he want his face to be on the TV giving an explanation for the fall of this man? What would it do for his business? How was he going to face those who had said all along that the man was a fake? For some of them there would be little cost to being in this circle, but it could affect his livelihood. Gene was having the same thoughts as they prepared for the public.

Shouldn't one of the Pastors make the statement? After all they were very good at gently explaining the offence of sin. They could handle the fall out better than anyone. Asa was known in the community as a rock in the midst of crisis. Pastor George could draw himself to full height and defy any tough questions. The press was accustomed to their polite non-answers and would allow them space as public figures.

"Luke," George sat as he spoke. "What are the terms of the agreement you men have worked out?"

This is good, thought Luke, try it out on George. If it makes it past him then I should be all right. "First, I am going to affirm our love for Pastor Greene. We have arranged for him to submit to a small group of local clergy for the purpose of restoration, we will continue his salary and benefits for one year and then reassess the situation. We realize this is a blow to the life of the church, and to the community, but ask those who believe to continue in prayer that healing can be rapid and complete."

"Then, I am not going to take specific questions as the investigation is on going and we would not want to misinform." He looked for approval.

"Asa, what do you think of it?" George's reply was a bit disconcerting. They had never heard one Christian leader defer to another on such an important statement. Gene leaned forward as the dynamic of the "Life Circle" began to flow.

"I think we should say he did it." Asa was to the point. "That is the question on every one's mind, 'Did he do it?'"

"Yes," the tortured voice came from the corner chair. "Yes, I did it, and I have been doing it, and unless something happens, I will do it again. I have always answered for myself and I do not need Luke to stand in the gap for me. I can take my own shame."

"And talk your own game." She stood in the door way coffee in hand. "You have always controlled the game, Robert Greene, but this time you will not. Luke will talk or I will walk. We are going to stand with you, but not under or behind you. We are going to face the reproach of this side by side, but you are not going to sweet talk your way out of it." Her voice did not break. Her stand was firm. Her resolve, unwavering.

"Good, then," said Asa, "I believe the press has assembled in your flower beds. Let's see, George and I will stand on the ends and you four men in the middle. Bob, stay out of the picture. Lisa, you look great in the kitchen. Please do alert the kids to not come through the door."

As they approached the door George turned to Ernie, the

fourth Board member. "Do you always say so much?"

"Usually." Was Ernie's terse response and they opened the door to the scornful faces of those who were ready, instruments in hand, to toast a man who had fallen.

## The Forgiving Power of Prayer

Alone at last, or was it at last alone? Their eyes met across the room. "You knew?"

She turned and walked into the kitchen. She was not ready for this. Emotion flooded through her. Yes, forgive; but, how to overcome the desire to run out the back door. Wouldn't that give the press a chance for headlines?

He didn't follow. He was too busy with the three children staring at him from their hallway hideaway. They had heard it all. Wasn't that the very thing that had driven him to the streets in the first place? He had no place to hide from these four people. They knew his every thought. They governed his every move. They encircled him like a wall. His only doorway out was to violate their trust and here they were again.

They didn't say one word. They just stood there. He had no compulsion to reach out to them. He stared back. The silence brought Lisa from the kitchen. They ran to her, clutching her legs and looking back at him. What was going to happen? Was this the end of Mommy and Daddy? How was their world going to be put back together?

Lost, Robert Greene sank back into the chair's warm refuge. He didn't care what they thought of him. He just didn't care. He couldn't deceive four. They knew his heart. From the chair's embrace he looked in cold defiance. Who were they to judge him?

Who were these people? They looked like strangers.

There it was again. She had it again. The poise, the strength, where did it come from? How could she stand, towel in hand, children surrounding her, porch filled with press, a woman scorned, and still in complete control? How could she do it?

"Children," resolve resonated, "your father has committed a great sin. He has hurt God. He has hurt us. We will not hurt him back. We will forgive. You must pray for him every day and trust the Lord to touch his heart."

"And what if we don't want to?" So, even this group had a spokesman.

"'Want' is not the word," she declared. "'Forgive' is the word we will use. We will forgive him and pray for him every day. We are going to work with the men outside and make a Circle of Life around your father. We are going to be like a life preserver to rescue him from a sea of sin. Each of us will have to do our part and we will succeed."

"Mommy," Megan stretched to make herself heard. "What is going to be my part?"

Lifting her from the floor, and clutching her to her heart, Lisa smoothed her hair and said, "You pray just like we do on the lonely nights. You pray for this man. You ask Jesus to put a shield around him. You ask Jesus to forgive him. You ask Jesus to reveal his sins. You ask Jesus to give him the strength to not sin anymore."

As he watched them look into each other's eyes, Pastor Robert Greene got the message. It was through their prayers that he had been brought to this point. Painful as it was, he began to feel life in his soul for the first time in years. Megan had been praying for him through the darkest hours of their lives. Lisa had taken her pain to the prayer-closet and the events of the day were answers to her intercession.

Six men came through the front door and the "Life Circle" was formed. It had been a difficult fall from Pyramid's Peak. He did not know how he would ever unlearn the old ways and learn the new, but in that instant he felt their love. Not their loyalty, not their undaunted support, but their love touched his heart. Tough love, but love indeed. he was ready to go on in God. He was ready to grow.

*Chapter 11*

# Developing Disciplines

**THE missing "D" in Pastor Robert Greene's life is Discipline.** His "Life Circle" is going to have to develop guidelines for his life. He has lost the disciplines necessary to continue as a Christian leader. His wife will play a greater role in this restoration than any of the others. She will be his restraint as grace changes his heart.

In Latin America, as the revival fires began there were many Pastors and leaders who lived an undisciplined lifestyle. Left in charge by a departing generation of missionaries, they established little kingdoms. The church was surrounded by civil war. Thousands were killed and the country became a battleground for ethnic and ideological struggle. Nowhere was the division more apparent than in the church. By 1974 there were 162 registered Protestant denominations in the capital. Catholics were being fed a "Liberation Theology" based in land reform and the Evangelicals fed on a steady diet of "Hate Catholics Because They Hate You."

The ideological landscape looked like the Nile Valley with pyramids built in every direction.

In 1976, the Lord rocked the place with a very strong series of earthquakes. These pyramids fell, and from the ruins have come great revival. One of the early signs of revival was the placing of former leaders "Under Discipline" for the sinful practices in their lives. Many of those who have emerged as great leaders in today's world-reaching revivals were at one time under discipline in the church. It is this same sojourn on which Robert Greene is about to embark.

## Word Discipline

**The Word of God is a light unto our path and a lamp unto our feet. When consumed on a daily basis, it lights the "Big Picture" and also the small details of our lives. It teaches us who God is and who we are in the light of His being. It has been from the beginning and it will remain when all else is gone. It is the heart of our first discipline.**

The power for life is in the Word. For by it all things consist. The Word was made flesh and dwelt among man. In His time of testing, Jesus used the written Word to dispel the lies of the devil. Robert Greene was unable to call upon the power of the Word of God to deliver him. From the pyramid's pinnacle, he could no longer remember the basic discipline of Christian victory, "Your Word I have hidden in my heart, that I might not sin against You" (Psalms 119:11).

To get to where you are going, you must have a Word discipline. Your "Life Circle" will help you develop this discipline. While preparing for ministry, I had the joy of attending a church-based Bible School. Sanctioned by the Assemblies of God, it was a place in which the students were expected to study every book of the Bible in detail. Having left a successful business to follow the Lord's call in my life, I was older than most of the students. However, there were two men in the class who, because of the paths of their lives, had also been delayed in entering this discipline environment.

## Fellowship in the Word

The three of us became a "Life Circle." We enrolled for 25 hours per quarter, which was much more than a full academic load and went to work on the material and on each other. Alan, one of the men, had been in prison when he received Christ. The hours spent there had become Library time for him. He had read the Bible through several times and had amassed a vast reservoir of memory verses.

Sam had come through Teen Challenge and had replaced a drug addiction with a Word addiction. He too was a walking collection of memory verses. He loved to ask, "Where is that in the Word?"

As we progressed in the first term, I began to realize that my previous study, growing up in a Christian home, and always being in the life of the church, would have been sufficient for the completion of the course material. However, I found myself using every tool I could to get an edge on understanding the Word. Why would I stay up until 2 AM night after night and leave early for class the next morning? I did not want Alan and Sam to get ahead of me in the Word of God.

We were often more completely prepared for a class than the teacher. They would make a good presentation and we would look at each other's pages of notes which had more depth on the verses. After classes we would sit together and share what we had found the night before. Often our instructors would join the group. The standard of the school was raised because of the informal "Life Circle." After we studied Greek and Hebrew, our hunger for the Word became insatiable. We added evening teachings to the schedule and would sit for hours and explore individual words in Biblical context. The fellowship, while a bit competitive, was wonderful as the Lord led us through His Word.

It was this fellowship that kept us going. After about a year, our wives expressed an interest in knowing what had happened to the three of us so we invited them to a Tuesday night class we took from a very demanding instructor. We cautioned them that this man would really put you down if you came unprepared. They audited the class and it was wonderful. We wept together at the beauty of the Word and would fellowship into the wee hours over the verses we had learned.

As we finished the course and followed the Lord in ministry, we were miles apart. Alan went to work for Charles Colson Prison Ministries, Sam began to pastor a church in Pennsylvania and we moved to Latin America as missionaries. We missed the fellowship. Our "Life Circle" carried on only in the memory of the way the brothers would challenge each other about a particular verse. Even now, more than 20 years later, I can see their faces and hear their voices each time I misquote a Scripture or Scripture address.

Time in ministry moves men of God away from such fellowship in the Word. We rarely find someone or a group of people with whom we can get together regularly to study and with whom we can fellowship in the Word. After our "East Coast Pyramid" came apart, I sought out some men of God and our very simple Thursday breakfast became a "Life Circle" for up to 50 ministers. So much of our Word time had become sermon preparation. Our search in the Word was to find food for others, but when crisis came we found we did not have strength for ourselves. We determined that the breakfast conversation would be Bible based. You would have enjoyed listening to any of the tables of four as plates were pushed aside and Bibles came out.

**"Verses Only" was the rule and "Let the Bible Interpret the Bible" the standard** as these men sharpened each other in the tool of their trade. Presbyterian and Baptist read the Word on Baptism. Anglican and Assemblies analyzed the Cross of Christ. And all the while they loved each other. Relationships were formed at those tables. Trust was built and over time they unified in prayer for their cities.

Robert Greene had avoided such a group. He was the authority on the Word of God in his world. No one else could speak into it. He had learned the air of distinction that kept him above the crowd, but his message had long ago taken a path divergent from his life. Because he had rehearsed the mantra for so long, he had begun to believe it himself, "When I am in the pulpit, God is speaking through me."

Lisa had been the first to see the difference. The Bible had been Robert's constant companion in the early years. His conversation had been filled with its light and his spirit kept strong by its absolute declarations. She had enjoyed so very much his little sermons to her each time she had a need. He would quote the verses with ease and apply them in their lives. Each time he was to speak, he would draw away and read the Bible and pray for a fresh word for the people. How he had worked at the messages.

After the church reached two hundred, he had begun to teach on themes as opposed to expository teaching. It was necessary she supposed, but the zeal for study wasn't there as it had been. When she had asked him about it, he just grunted something about all of his time going into the people and not having any left for study. She noticed that he no longer carried his Bible as he had in the past and sometimes actually left it at home for several days at a time.

Lisa had been concerned and had started to leave study guides in his path. Each one he had picked up and tossed aside. How was she going to bring him back to the discipline of the Word? She had invited a few friends to their home for dinner one evening. People with whom they had gone to school, she thought that they could play a little game of Bible Quiz after dinner, but Robert insisted that the men wanted to watch basketball and the women were relegated to the living room to chat about children. She had become deeply concerned at that point and had decided to ask him why he no longer loved the Word of God.

A Wednesday evening service had provided the perfect opportunity. Robert could not find a verse. Teaching from memory instead of notes, he had become so flustered that he had to stop talking, retrieved his Bible from the chair behind him, and stumble through the pages searching for the location of the verse. It had been Gene, the Dentist, who had called out the reference to the embarrassed Pastor.

Things had been very quiet in the car on the way home. She had waited for him to open the conversation. He had been in no hurry; but, realizing the need to say something, he had laughed that the dreaded "It" had happened. All the years of teaching the Word had filled him with so much knowledge that the verses had overlapped in his mind.

"I think you didn't study today," she had said with conditioned gentleness. "In fact, I think you didn't study this week."

"What do you mean?" he had responded. "I know those

Scriptures by heart, I can go for awhile on accumulated knowledge. No one there would know the difference."

"Well," she had ventured further into the vein. "I was there, and I know the difference. Your Bible is not doing either of us any good under the nightstand where I put it a week ago. And anyone who has been with you for more than five years is beginning to recognize the messages. I think it is time you got back to the Bible."

Silence isn't golden when it is aimed at you. He had been furious that she would say such a thing to him. After all, hadn't he been working 80 hours a week to build the church? All she had to do was sit in the front and stay out of conflicts and he would guide the ship. Now she had said that he was not skilled enough in the Word to lead the people. Slamming the car door behind them, he had nearly torn the kitchen door from its hinges. He had raged to the bedroom and reached down to retrieve his Bible from beneath the nightstand, but had not been able to find it. When she had caught up to him, he had been on his knees searching furiously for the lost book.

"I moved it to your throne behind the pulpit," she had said. "Remember, it was right there when you needed it?" She had stood firm waiting for the rage to rise from the floor and burst forth in her face.

He had slumped forward in resignation. "You are right," he had said. "And I am sorry for the way I have behaved. Lisa, there is something blocking me from the study of the Word. My Bible used to feel perfect in my hand. The Lord used to speak to me from the pages. Every day was an adventure in it. Now it is dry and removed from my mind. I do not even want to talk about it."

**Robert Greene had lost the discipline that had made him Pastor Greene.** Because of his pyramid structure, only Lisa could crash into his world. Had they been a little more aware of the horrific results of this loss, they may have averted the nightmare they would face, for the Bible was not the only precious thing that no longer felt at home in his hands. Robert was showing the early signs of decaying morality. Once the discipline was gone, the light would dim and worldly rationalization would take the place of the absolute truths of the Word of God.

## Maintaining Word Discipline

**There are several ways to maintain a Word discipline.** Some try the Read-the-Bible-through-in-a-Year Plan. In this way, the entire congregation is reading together and it is hoped that they will converse in what they have read. Sunday school participation in which assignments are given has proven to be a good way for a smaller "Life Circle" to sustain one another.

Weekly studies like those produced through Promise Keepers are very good with their question-and-answer discussions which keep us reading. Many of the cell-based churches use questions from the Pastor's Sunday sermon as a guide to weekly reading.

All of these are the expression of a "Life Circle" sustaining them through mutual participation in an overall plan.

Many in the restricted churches of the Communist world do not have the luxury of frequent meetings. They have adapted to personal study. Psalms and Proverbs can be read through in a month. New Testament and Gospel passages are often related in daily living in developing countries. In every place where the Bible is found, there are innovative ways to study its pages.

However, there is no substitute for the individual taking up the Bible and studying it simply because we are instructed to do so. Success is the measure of the distance between desire and discipline. To be a disciple we must leave the level of desires and attain the level of discipline. The successful person is the one who excels at the things he does not want to do.

There is an issue at hand when we attempt a Word Discipline. It is the second of the forces we mentioned before: "The Right to Speak." When we submit to a Word Discipline, we allow the absolutes of the Word of God to cut across the flow of justification and rationalization in our lives. The Word is written in very simple language. It is given to cause us to prosper. It contains over 6,500 promises to those who love the Lord.

Why would we not pick it up and read it? We want to be our own God. At the end of the day, we want to rule ourselves, make our own decisions, and go our own way. We struggle with power and control. Robert Greene did. Leaving the light of the Word, he crashed into a dark ditch.

There are times when the Word is very "dry." It is intended to be so. It is designed to dry us out so that we thirst for the Spirit of God to move on those pages and give us refreshment. Jesus was very dry after 40 days without food and water, but when temptation came, he spoke the Word to wash it away. You may go through dry times, but these are in the plan of God to bring you close to Him.

It takes time to read the Word. Yes, but not at all as long a time as it took to write it. What is your morning devotion? Everyone has one. Some people rush to their computer to see how the overseas markets did. Others check their emails to see if their friends are still their friends. Some others fellowship with the false gods of caffeine and nicotine. And there are those of us whose worlds are ruled by the needs of our pets.

There are times when we do not understand what we read. This is a time to stop and pray that the Lord will give understanding.

He is causing us to dig a little deeper. I like to use these times to read a commentary from a 19th-century writer to see if what I am understanding today is consistent with the volume of the teaching of the church.

As we read we often think of others in the "Life Circle" and can call them and encourage them from the Word. I really enjoy the calls and emails I get from Bob. And now I understand that his Pastor is taking his calls and rejoicing in the Word as well.

Robert Greene will be restored to his Word discipline. The new "Life Circle" will see to it because they understand that the best way to fall into darkness is to turn off the light. Lisa will play a big part, proving that the Word is true, "It is not good for man to be alone."

*Chapter 12*

# Worship Discipline

**THE second essential discipline of the discipleship process is Worship.** Robert Greene once had a vision. That vision was shared with his wife. They joined to themselves others who would embrace the vision. That group had been a "Life Circle." As the group had grown beyond his ability to control it, he had departed from the basic disciplines of discipleship. He had become the lord of the work.

At first he led all of the worship. There was no one else there to do it. But soon gifted people came into the church and he was able to delegate that ministry. Unfortunately, as he did, he forsook his worship discipline as well.

## Creating a Sanctuary

**Biblical Discipleship is a process with a product.** Jesus spoke to these whom he called to "Follow after Him..." He promised them that they would realize a change and become something other than what they were at the time of the calling. They would become "Fishers of Men." This would be their title in the Kingdom to Come.

The discipline of worship creates a sanctuary for the Lord to inhabit the praises of His people. It is in this sweet sanctuary of worship that we learn the voice of God for our lives. Like so many others, Robert left that sanctuary and slowly crossed to the shadows of the other side of the street. Bringing him back will require a restoration of the worship discipline.

You are going on in the Lord. Your "Life Circle" has been formed. You have a vision for what the Lord wants you to do. You are bringing together the resources necessary to accomplish the vision the Lord has given and you are excited and ready to become a Fisher of Men. As you establish these disciplines in your life, you will become the product of the process.

**Ephesians 5:18,19 teaches us to "…not be drunk with wine, in which is dissipation; but be filled with the Spirit, speaking to one another in psalms and hymns and spiritual songs, singing and making melody in your heart to the Lord,…"** This is the worship discipline in its purest form. Through the practice of the points of this Scripture, we will see the release of "Transformation Power" in our lives.

## Releasing Transformation Power

**The first point is to flee the former excesses and worldly ways.** These are ways that will choke out the promises of the Word of God in our lives. You may know people who received the Word with gladness, but have also received a good dose of weed seed and who are not doing well in Christ's discipleship plan. They received Jesus, but they assigned Him a place in their lives rather than allowing Him to become Lord and Master of all.

**The worship discipline teaches us to ascribe value to the Lord and His ways.** In teaching our hearts to see the value of Jesus and His people, we divorce ourselves from the ways of the world. Each time we ascribe worth to eternal values, we destroy the altars of temporality in our lives.

**This also opens our hearts to the second point of the verse, "But be filled with the Spirit."** This is that same Spirit that was moving over the face of the deep when God spoke and the worlds were created. It is the Word and Spirit combination which releases the creative power of the Lord. It was this combination that formed all things. God spoke into an environment bathed in the Holy Spirit. So it is in your "Life Circle" and in your heart. When you worship, you bathe your heart in the Holy Spirit. God can then speak into that heart and create your vision. He can send the power that will transform you to a Fisher of Men and will make your "Life Circle" Fishers of Men.

Much has been said of the infilling of the Holy Spirit. He replaces the spirit of the world, the prince of the air, the god of this world in His ability to fill our vessel. When we realize that the Holy Spirit has come to live in our hearts, our "world view" and personal perception are changed. Is He different from Christ? No, He is the Comforter described in John 16.

In John 16:7 Jesus tells the Disciples that He will send "another comforter". There are two words in Greek for "another", "heteros" and "allos." "Heteros" is defined as, "Another of a different kind." "Allos" is defined as, "Another, which is in substance and manner the same as the first." Heterosexual relationships are between different genders.

The word used to convey the meaning of Christ was "allon parakleton." In other words, the Holy Spirit is Christ in a new form. He is Christ with Omnipresence restored. He is that Christ in you that gives you the hope of glory. He is the King of all Kings and the Lord of all Lords. You are a mobile tabernacle of the Holy Spirit. As you worship, He makes Himself real to you.

## Releasing Creative Discipline

**The third point is what Jim Derrken called "Soul Talk." We talk to ourselves. It is true. We talk to ourselves in psalms, hymns, and spiritual songs.** The melody of your worship discipline releases the creative power of God through you.

The late James Derrken wrote a little book called "Soul Talk." From the early '70s Brother Derrken ministered to the Jesus People in Northern California. This group was composed of a bunch of hippies who by personal account had no discipline. They had drifted to California in what the youth of America then called, "The Age of Aquarius." It was a time of rebellion to social norms and the declaration of self. Jim had received from the Lord a very special call to these young people.

I first encountered those who he was discipling in Guatemala. They were a very disciplined bunch who moved about in a tight circle. Decisions were referred to the group, which at the time seemed crazy to me. I had just come from Bible School and an independent Charismatic Church. Our styles conflicted, but there was a certain life-giving element to these folk and we were all the same age. The worship was fantastic.

We began to attend their Sunday worship and through the week I found myself being embraced by their members. A very loving bunch, I found many of my walls coming down. I now realize that they were enveloping me in a "Life Circle" and the walls were those of the pyramid from which I wanted to rule. Their sense of selflessness and group identification was very appealing and even as I write I can recall the way they kept their soul life in control through worship discipline.

Our soulish realm will get us into trouble every time. It is made up of our will, intellect, and emotions. They had been taught to control the will through songs of surrender. Musical by nature, they had learned to play flutes and guitars as they drifted. Naturally poetic, they gave themselves to the song of the Lord and often we would just sing and compose lyrics as they came to our surrendered intellect. From them I learned the discipline of always having a song on the ready to sing to myself when I could feel the heat of emotion rising.

We learned how to speak to ourselves, reminding ourselves of our smallness and the greatness of the Lord. When we needed to love more, we sang of His great love for us. When we needed to think clearly, we sang the songs of David. Often we just sat and tried to put a melody to the Scripture verse. The fellowship was very sweet in the most adverse conditions. The Lord was indeed filling us with His Spirit.

One of those men, Francis Anfuso, re-emerged in my life just a while ago. We were brought back together and the first thing I asked him for was a copy of "Exploits" a tape he had produced with Ern Baxter and Bro. Derrken. I had given out over a hundred of these to pastors and missionaries. It was the most successful motivational tool to World Harvest I have ever heard. On it were songs which Francis had written. They were so compelling. Every person to whom I gave a tape responded with a desire for more.

We traveled together for a few days. Our prayer teams mingled and one morning we worshipped together in a hotel conference room. He was not present at the start of the meeting. A young lady led the worship but, as I closed my eyes, I could feel the same anointing that had been there so many years before. Soon I heard his voice beside me and, when worship finished, he introduced me to his daughter. The discipline had passed to the next generation and with it the ability to speak not only to her own soul; but also to the souls of others.

Members of that first group of former Hippies are now bringing forth fruit around the world. They became Fishers of Men and they have produced the same from Guatemala to Beijing, from Miami to Moscow. The Worship Discipline was their call and they followed.

When we met, the most unnerving thing about them was their approach to hearing from the Lord. Our group was still under the influence of the tapes we had bought into on retreat. Even though I had relocated to Latin America with several families from our church, we were not a "Life Circle." My wife and I handled the money and accounting and the group looked to us to make the decisions. We were, by mutual agreement, in a very precarious position at the top of a pyramid.

As I tried to hear the leading of the Lord for the group, my own needs became interference to His voice. We did not worship together for a number of reasons, and I should have seen the trouble coming. I had just come from a Word Discipline that was a great joy and several of these folk could not read. The contradictions in our lives were extreme. The group was fraught with tension. It was not a pleasant situation, but to their credit they accomplished much of what they had set out to do.

Carlos, Jim, Dick, and Bob, the leaders of the other group were a "Life Circle." They shared decisions and sought the Lord together. I watched it. Each time I would ask one of them a question, the person would respond, "Let me take that to the group and get back to you." Now that was unnerving to a young "Moses." I finally struck out at them and our group in total frustration. It wasn't the people, it was the pyramid that created a very difficult time.

## Communicating With God

**"Soul Talk" is summed up in the Scripture, "Bless the Lord oh my soul and all that is within me bless His holy name."** When we have a worship discipline, the voice of our own ambition is silenced and the still small voice of God can be heard.

**Worship is a dialogue.** The two voices involved are our's, and the Lord's. **We bring a sacrifice of praise to Him.** This is voluntary. It starts with your decision to praise Him regardless of the circumstance. This sacrifice of praise takes the place of complaint. As you choose to sing praises to the Lord, you position your spirit to receive His grace. You need a blessing. It could come in many forms. God will determine what is best for you and will send it forth by His Word into your spirit.

**He responds.** He gives you a sense of His love for you. You can feel it in your soul. It is a warmth or certainty that the Lord has heard your cry. An expectation begins to arise and your song ascends into a more worshipful level. You are thinking of His abilities instead of your need. His great value has overshadowed the value of your desires. This is His release to worship.

**You respond.** Now you are open to "Spiritual Songs," as the melody in your heart wants to give expression. Your will is aligned, your intellect is beginning to think on the Lord and not on self. Your emotions may catch you up into ecstatic expression before the Lord. This is a good thing for you are speaking to yourself in psalms, hymns, and spiritual songs. The melody in your heart is finding expression through your soul.

But, the dialogue has just begun for now your worship discipline has taken you far beyond the limits of your intellect, will and emotions. You are now in a position to receive what the Lord has for you. You as an individual or your "Life Circle" can now hear from the Lord. **Wait before Him.** Sing softly if you must. Be still if you can for He will give you direction. He will speak through each person in the way in which He is discipling them in this discipline.

When He has ministered to the group, do not rush off from His presence. Often, He will wait to see if you will give thanks. Often we do not. If we have a bit of the pyramid alive in our souls, the next voice will be our own giving direction. If we live in the midst of a "Life Circle," the next voice will be that of another as we have our soul in check. Dietrich Bonnhoeffer in his book *Life Together* took this to the point that the one who broke away from the unity of worship would be the one who would turn on the group in times of persecution. His point is that the discipline of worship brings humility, which will no longer exert individual power.

Pastor Dowdy has found the same truth in her business fellowship. Pastor Jon did not have it. It is easy to see if you do. The next time you meet with your ministry team, try the dialogue of worship and see what happens. Worship is not the few songs we sing before the leader speaks. It is a discipline, which can grow into a love affair with Jesus.

**The worship discipline teaches us the voice of God in our lives.** It is that discipline through which prophetic anointing flows. It is the discipline which develops our prayer life. It is the discipline, which develops humility. It is the discipline which develops selflessness.

Selflessness demonstrated over time equals trust, and all successful living is built upon this rock solid foundation. Marriage must have trust. Business must have trust. The global community must have trust. The church above all must have trust. It is the worship discipline that develops the essence of trust. You can trust a person who daily comes before God to sing and to praise Him. A person whose first words in the day are praises and who humbles himself before the One who has given all for him.

*Chapter 13*

# Work Discipline

**THE third necessary discipline to get from A to B is the Work Discipline.** This is something Pastor Naomi Dowdy relishes. Work is defined as moving one gram one centimeter. It is measured in Ergs. This seems like a good word for it as there is a certain amount of strain involved. The work of bringing Robert Greene back to the ministry is going to require a great deal of what follows in this chapter. Here, you are going to find five planks for building the deck of your BRIDGE to ministry.

The letter to the Galatians is a study in the contrast of the flesh and the Spirit. Having begun with a vision that was born in a "Life Circle," do not stray to a pyramid for its implementation. Your Word discipline gave you light for your path. Your worship discipline taught you to hear the guiding voice of the Lord as you go. Your work discipline will bring you through to completion. You have begun well, press on to the goal.

Galatians 6:1–5 will be the undergirding Word upon which the "Life Circle" will set the pathway of planks which will get Robert Greene to his new destination.

## Forgiveness

**"Brethren, if a man is overtaken in any trespass, you who are spiritual restore such a one in a spirit of gentleness, considering yourself lest you also be tempted" (Galatians 6:1).**

This first plank is so very hard to do. It is going to take a lot of work for you to see the trespasses of others and still be able to embrace them. How wise of the Lord to teach us to pray, saying, "Forgive us our trespasses as we forgive those who trespass against us." To the measure we forgive, so shall we be forgiven. Does that mean that Lisa Greene is not dying a thousand deaths? Does it mean that Luke, Gene, and Bill, are not dying to take at least a verbal shot at Robert? Of course the hurt is there. So is the burden of sin, but this burden will have to be lifted and set at the cross, or none of them will get to the other side of the experience. **The first work of this discipline is the labor of forgiveness.**

Where will Lisa Greene get the spirit of gentleness necessary to walk through this process with her husband? He had noticed her regal nature, the quiet strength that she had gained in hours spent in the prayer closet with Megan. She had learned to control her soul in his absence, but now she would need greater control in his presence. Through the nights she could think of him as a victim. She had agonized through hours of self-analysis. In the light of day she would see the corruption of his spirit which was the result of his abandoning Christian discipline.

**Those who have fallen are aware of their sin.** They may try to justify, but even that justification tells us that they are aware. Robert's realization that Lisa knew all along that he was in sin was the power that started his transformation. She holds the key to his future and has the most work to do. The children will look to her for direction. She will have to model the gentleness that one day will keep them in times of disappointment. She will draw very deeply from her word and worship disciplines for the strength for their transformations.

Self-consideration is an outgrowth or fruit of our worship discipline. We see ourselves as small before the Lord, but now we must see ourselves with the same smallness in comparison to our brothers. And not just the nice ones, but especially the ones who are not nice. We must be smaller than those who are in sin. This is work. It requires discipline. Something heavier than the burden of sin has to be carried away. The burden of pride must be carried to the cross before this plank can be fully laid.

Every good work is tested in relationship. Hebrews 6:12 teaches us that we are to "Be not slothful, but followers of them who through faith and patience do inherit the promises of God." This word "Patient" is used in interpersonal relationships in each NT context. Imagine that the Lord gives us a promised destination, and we are expected to get along well with the other people going together. This is work. This requires discipline.

## Carrying Another's Burdens

**"Bear one another's burdens, and so fulfill the law of Christ" (Galatians 6:2).**

**The second plank is as heavy as the first.** The weight of Robert Greene's sin is going to be carried by his "Life Circle." He is very fortunate, because these people are willing to make the effort. They could have said, "No thank you." Asa could have kept on going. But he carries the burden of the church in the community. He came down from the pyramid years ago and emptied his hands of self-desire and disciplined himself in a small group to carry the burdens of the others. Not every one will be so willing. Each will count the personal cost of identification with the sinner.

Jesus counted that cost. The cross was heavy, but the sin was crushing. He encouraged each of us to count the cost. The more effective we want to be the more expensive it is. The elders had to count the cost of covering Jon. They were not willing to do it. His move to a bank for funding created a financial burden they were not ready to bear. Him doing so without talking it over with them was the last weight that broke their relationships. Selfishness broke the back of the Trust Formula.

**"Life Circles" are designed to humbly carry others.** My Mother's circle carried many families as they came through economic difficulties. I know there were times when she would have uncovered someone's sin, but she would purse her lips and love again. How I appreciated her depth of love when, in my rebellion years, she sang the house to sleep with her worship discipline.

**This second plank has splinters.** You be careful how you pick up the boards of burden. The cross on which the Lord carried you was rough. It dug into His flesh and the sins of others will dig into yours. Ministry is not meant for the faint-hearted. There are very few who can live in the middle, being the servant to the gifts of those with whom they do not always sympathize. You will get stretched in places you do not even know. The splinters will get in your fingers and your eyes. Your grip on things will be affected as will your ability to see clearly. There will be tears of sorrow and joy. You will find times of exhaustion as the burdens of others are brought to you.

There are two gloves you can wear to govern your actions, which will prevent splinters. Each is a product of a developed work discipline. **The first is the glove of the right hand,**

**having your own work done.**

**Success is the measure of the distance between intention and production.** We live in an age where "intention" is almost on an even level with "production." If someone was supposed to call and they thought they would, but they didn't then you are to excuse the fact that you lived without information until they finally get enough discipline going to call. I have lived on both sides of this glove.

It was so very easy to be critical of others in ministry until my name became the one at the bottom line of the page. Oh, the transformation power of the realization that I could point to no one other than myself. Now as the South East Asia Prayer Center takes shape and budget and responsibility increase, I realize much more clearly the actions of some of my fellow ministers. This allows me to carry them in prayer and giving, and destroys any desire to criticize them for the way they do things. Eternity will record how many splinters have been stopped by the first glove.

**The left-hand glove is "motive control".** Why am I doing what I am doing? In our Asian culture, the left hand is the toilet hand. We never greet anyone with it. We never try to serve anyone anything with it. It is placed under the right arm when gifts are extended and received. Jesus said to not let it know what the right hand is doing.

The right hand is the "giving hand" the left hand is the "getting hand." I must cover the left hand with the glove of motive control. I must not enter into service with any opportunity to get. My motive has to be pure for the Lord to reach through me to restore a person to right relationship with him.

Picking up the second plank will be the task of both Pastors in Robert Greene's "Life Circle" for they can both get splinters from being associated with him. They could also both gain members with the information they are about to receive. His time of testing has proven him to be wanting in righteousness. Their time of walking with him will also test theirs. This plank must be lifted most carefully.

## Humility

The third plank of the bridge of work discipline is central to spiritual success.

**"For if anyone thinks himself to be something, when he is nothing, he deceives himself" (Galatians 6:3).**

Pyramid perching is the pinnacle of pompous personal perception. Pastor Greene's biggest problem was that he really believed the people came from so far to hear him. While he was very good, the greater dynamic that he had was first the presence of God and second the loving nature of the church as demonstrated by its members. One of Pastor Dowdy's strengths is the realization of the value of those around her. She is a master at "Life Circle" development and is always including others in the circle.

**Self-elevation leads to swift deflation.** To be able to put this central plank in place, you will need the help of others. It is very dense, will hold the bridge together, is extremely heavy, and will require you to release your end when the other brother releases his. If you can see yourself as less than the least of these, you can succeed with this plank. If one ounce of pride rises up in you, you will hear yourself make the sound of work. The Ergs will get you.

Personal responsibility teaches humility. Each member has a ministry and as a result learns humility. Be careful when you are doing well for a man's praise is his testing. There is no replacement for humility, it is the foundation stone of successful ministry.

## Giving God the Glory

The fourth plank is wonderfully clear.

"Let each man examine his own work, and then he will have rejoicing in himself alone, and not in another" (Galatians 6:4).

From the word discipline we see the perfection of Christ. Jesus did not half-heal. He did not half-deliver. He did not tell two thousand to come back next week. He did not control the wind and leave the waves. He did not leave Lazarus in the grave clothes. What He began, He also finished. In John 17 we can read His report to the Father.

Because He is the embodiment of the mission, Jesus tasted the completion of His task before it had been accomplished. As He prays here we see that He is reestablishing that relationship which he shared with the Father before coming to earth. The dialogue of the Godhead found here is unique to John. Jesus is counting the accomplishment and the Father reaffirms the waiting reward. It is this future hope that will enable Christ to endure the cross, despising the shame, but looking forward to the joy that is set before Him. Even as the disciples struggled through their night of terror, Christ was secure for He could see that beneath the load of sin, He was still the Son of God.

Lifted up on the cross, He could view the garden tomb. He could see through the pain to the victory. He could set His soul on the promise of the Father having seen himself raised in a figure. You must see through the pain. You must examine yourself to see if you are serving the Lord or self. You are the only one who in that prayer closet with the Father can determine if you have done all He wants you to do.

If you have not, get with it. If you have, give Him the glory. And rejoice with Him in it. When others perceive you to be something very special, realize that they are seeing Christ in you. Give Him the glory and should He whisper a "congratulations" in your spirit for a job well done, keep it to yourself. You will need it the next time you fall short of a task or promise.

## Doing Your Part

"Let each one bear his own load."

The fifth and final plank of the work discipline is the heaviest and most awkward. You are not going to have help carrying it. Your work will be seen in the public arena. The people waiting to cross your BRIDGE will be a pressure as they either cheer you on or smother your zeal. You have to pick it up and go with it.

Those who do not carry their own weight in the "Life Circle" will eventually become critical. They will cause a cramp or get out of joint. They are Bonhoeffer's "ones to be watched." So often we commit to a ministry and then do not carry it as our own.

Every nursery superintendent knows the Sunday AM panic when the phone starts to ring and well-intentioned folk explain why they cannot serve that day. This is not unique to the church. Once I noticed that a local store had an unusual number of senior citizens among the employees. I asked the manager if that company had a particular policy concerning such things. He was quick to respond that when he scheduled young people, he was faced with 50-percent absenteeism. With the older people, they honored their job and were faithful to the schedule.

Why do 80 percent of mission plans not reach their goals within the time allowed and within the budget? Because the people involved are waiting for someone else to carry the final plank. It is not going to happen. Ministry is 10-percent inspiration and 90-percent, perspiration. There are no short cuts.

Each member of the "Life Circle" must do his part. As you meet, be sure to draw participation for each one. Learn to listen to them. Set the example by having your handouts ready. Have your schedule cleared to prioritize for them. Do not take phone calls during the meeting. Prefer the rest of the group. As the central server, be prepared.

When I served a large, Assembly of God church in the United States, I was responsible for several areas of ministry. We formed "Life Circles" for each and, as the server, I set the format for our meetings. Three nights each week I participated in two meetings.

The format was straightforward. Knowing that time was tight, the people honored each other by being punctual. I prepared for them a handout having the theme of the meeting and guiding Scriptures. We started with about 15 minutes of praise and worship. Then I spoke for ten minutes from the Scripture text, making application in the context of the work we were doing. Then we heard from each of the members on developments in their areas of the ministry. With all that we still had 30 minutes left in our appointed meeting time.

As we had fulfilled the worship discipline, the word discipline, and the work discipline, we could relax and hear from the Lord. We would begin to pray over the things we had shared. Where problems had been mentioned we would ask the Lord to solve them. Where praise reports had been shared, we joined in the rejoicing. We encouraged each member to participate and enjoyed a relational free-flowing in the Spirit because we had a track to run on. No one voice dominated, but each person had the same opportunity for input.

Our scribe kept notes of all that was brought forth as a vision, or a Scripture, or in a prophetic sense. When the time had passed, we read the list to be certain that everyone who had said something was included.

The next morning, my secretary would type the record of the meeting and mail it to the "Life Circle" members. They had two weeks to read it over and pray about the next meeting. We stressed the importance of each of us carrying our own part to the best of our abilities.

One week I had an especially heavy work-load and failed to prepare the handout. I thought it would be all right as we had been meeting for almost a year on a biweekly basis and our fellowship was getting really strong. It was a disaster. Why? I had not carried my own part of the ministry. I had failed to prepare for the meeting. I had dropped the plank.

To some, my failure would not be as serious as Robert Greene's. However, to the Lord, sin is sin. The forfeiture of the work discipline is a disaster to ministry.

**The three essential disciplines of discipleship are the word discipline, the worship discipline and the work discipline.** These form a three-strand cord that will sustain you as you build from A to B. When one discipline is weak, the others will sustain for a season. When two are weak, the work will not develop. When three are weak, the fall is soon to happen. Robert Greene fell because he had abandoned the three essential disciplines. His restoration can occur only as they are rebuilt in his life.

*Chapter 14*

# Going On

PASTOR Jon, Pastor Dowdy, Robert Greene, and you have something in common. You all want to leave where you are and go somewhere else. Pastor Jon is on his way out of ministry. He has had as much of the pyramid as he can stand. He has "irreconcilable differences" with those he was once called to serve. Because he did not develop relational bonds with a "Life Circle," he was not able to sustain the work. He does not want to go on to "Life Circles," humbling himself before his elders, so he has little choice but to leave.

His open letter makes it quite clear that he sees no other way. He has splinters in his eyes and has lost the joy in his heart. He will, no doubt, eventually find a fellowship where he can be healed for the Lord will not forget his labor of love.

Pastor Dowdy is going on as well, having seen the joy of relationship in the business community; she has established the Global Business Network that is an international community of executives who want to pool resources to expand the Kingdom of God. She has managed to draw from the hearts and minds of these people the vision to reach the globe with the Good News of Jesus.

To do so, she has had to move away from a pyramid structure and into today's model of networking. She will never be the owner of the structure, only a servant to it.

Robert Greene has to decide which way he wants to go. He has to make a move in some direction. His flesh does not want to. He would like to sit still where he is. In fact, if it hadn't been for his former friends, the chief of police and the magistrate, he could have carried on undiscovered and this nightmare would never have happened. Instead, he is faced with a very grim decision. Will he leave the pyramid, or will he strike the rock?

## We Have to "Go"

**The "G" in our acrostic is the little word, "Go." Small but powerful, it is the element which determines success or failure in building a BRIDGE that will get your "Life Circle" from A to B.**

The spirit of man says, "Stay where you are." We resist change. In some ways this is a good thing. Establishing a stable environment for the family to grow is a good thing. Establishing a solid base financially is a good thing. Staying with your spouse is a good thing. Remaining committed to your children is a very good thing. Walking every day with Jesus is a wonderful thing.

However, to get from A to B we have to go, and that can be difficult. In this chapter we are going to look at five factors in the going process. You may be able to see more and I invite your input. I am constantly being faced with having to change location. For me it has become an unpleasant thing. Moving from country to country and locations within a country may seem romantic at the start, but after all these years, I look forward to settling.

# Jesus Is Our Stability

**"Permanence is stability."** There was a time when a trip to the city happened once in your life and was shrouded in mystery. If a young lady left town to live with relatives, it was assumed that she was with child. When she returned, she had a much wiser look in her eyes. Cities were born out of the need to establish permanent food and watering places for pack animals and then trains. Even in the developing nations, the Banyan or people tree with its water and wood is a tradition. Change is a very intimidating fact of life for us all.

When you fill out an application for employment, the first section is "Who are you?" The second section is your work history. The big question is, "Are you stable?" If you own a home, you are stable. If you have been employed at one place for more than five years, you are stable. If you are now living within 100 miles of the place of your birth, you are stable. Our entire world calls out for stability.

You would not want to be a "Roamer" or a "Drifter"; you want to have roots deep in family values. In fact, why would you ever leave the family business? How very unstable that would be. You definitely do not want to be a "church hopper" going about from place to place looking for a Pastor who will return your phone call.

Today we have virtual office, virtual business, and virtual corporations. We can, from a laptop computer and Internet connection, be anywhere in the world at the same time. Our ministry address is SEAPC@hotmail.com. We move about. I could be in Asia in the morning and America in the evening of the same calendar day.

Several years ago a lovely Christian couple in Singapore approached me. They asked me to hear them out on an idea they felt the Lord was speaking to them about. They wanted to create a website for an e-business that could be prepared to also answer questions about the coming change in millennia.

I suggested two things. First, I felt that the church was not ready to hear their vision. They would need to be ready to throw off the disappointment of finding out that we as a body are still dealing with mimeograph machines. The advances that have occurred in the past ten years are often seen as a threat and those who can understand and follow with them as unstable.

The second thing I suggested was that they form a small group of friends who would pray and believe with them for the creation of such a website and who could give them counsel in e-business development, something about which I know absolutely nothing.

We prayed together and I prayed for them often after, but our paths did not cross again until January 2000. I was at a Pastors' meeting in Malaysia when Barry came rushing up to me. "Have you heard?" he asked excitedly.

"Heard what?" I had prayed for him, but barely recognized the beaming face before me.

"We got the site up, we got the e-business going, you were right about the church, we had to carry it with a small group, but guess how many hits we had in December?"

"I have no idea." I responded, finally remembering what this was about.

**"Nine hundred thousand hits." He was excited. "Nine hundred thousand people got the Gospel in response to their questions about the change of millennia. Nine hundred thousand people in one month.**

Barry had moved on, and when I had a chance to share with the pastors in the meeting, I asked how many of them had preached to 900,000 people during December. In the course of his quest to move into the current age, Barry had struck on a very important understanding. Our permanence is in Jesus.

The spirit of man says, "Stay."

The Spirit of God says, "Go."

## Daring to Go

**"We need to first reach Jerusalem." "A prophet is not without honor, except in his own land."** Ok, for we weaker minds, which is it? Do we stay at home and try to overcome generations of memories to succeed or do we go forth into all the earth and make disciples of every nation? After the day of Pentecost, the Apostles stayed in Jerusalem. It was tough going. They were beaten and persecuted. The church grew and the Lord added souls. He had told them to go to the uttermost parts, but they enjoyed being where they were.

The pyramid was pretty well established with Peter at the top. Of course there would be factions who would back Mary for leadership and much later the "We follow Paul" and "We follow Apollos" groups would form. They liked it in Jerusalem. There was a sense of camaraderie even in the face of persecution.

About the time of Acts Chapter 8, a great persecution was brought against the church and they had to go out. Imagine that, the persecutors helped them to get going in the very thing Jesus had told them to do. They went out. They reached India, China, and even the barbarians in Europe. The gospel went out with them along the trade routes of the day. They began to multiply. They made an impact throughout Asia Minor. That is not too bad for a group of Galileans. Because they went, you and I have faith in Jesus today.

Imagine your life if Christianity was just a handfull of people huddled in Jerusalem. How would you have heard about Jesus? How could you know the joy of sins forgiven? No one else is going to take this message for you. You must go as Jesus did.

**"But don't we need to build a base?"** you ask. Sure we do. Ten people tithing will support one person at their average income. In most places around the world, that is more than enough. Do you need a 500-member base spending 90 percent of its resources to maintain itself so that it can send out one couple? Yes, we need to build a relational base, with a going mentality and a progressive vision.

## Am I Ready to Go?

**"But, I do not know if I am ready to go."** This third mindset is most restraining. How will you know? Will someone else tell you? Will you take a test to find out? How will you know it's time to make a move? What are your indicators?

Pastor Dan Livingston, a dear friend in ministry, said it this way, **"When the pain of where you are is greater than the perceived pain of where you are going, you will go. But, as long as the perceived pain of where you are going is greater than the pain of where you are, you will stay."**

Is that what happened in Acts Chapter 8? Did the pain drive them out to the nations? Was that the pillar of fire driving them to the cloud of presence?

One of the interesting things about change is it comes upon us. How often do we hear that someone's plans have changed? Mine usually do just after I have announced them. It is uncanny how we do not know the future as much as we would like to think we do. How can you remain calm and confident in the middle of change?

The Lord has promised that if we commit our plans to Him, He will direct our steps. He will not allow us to be ashamed. He has appointed good things for us. It is His delight to give us righteousness, peace, and joy. He will keep us in perfect peace, as our souls are stayed upon Him. If we delight ourselves in Him, He will bring our desires to pass. That is a lot of promises, but the one that I love for change is this, **"...if you diligently obey the voice of the Lord your God, to observe carefully all His commandments which I command you today,...these blessings shall come upon you and overtake you..." (Deuteronomy 28:1,2).**

I will never be ready to go. You will never be ready to go. Our three Pastor friends are not ready to go. If ready means that we can control completely the outcome of the going, none of us will ever be ready to go. That is one of the joys of a "Life Circle", we do not go alone, we have the multitude of counsel in which there is safety. We have learned to love one another so that we are free to make a mistake. We are surrounded by grace.

There is a children's rhyme we used to say, "One for the money. Two for the show. Three to get ready. And, away we go!" Unfortunately, the message has carried over to us as adults. We do not want to take a risk on going until we know we have the money and we know we will show well. We want to have every question answered before we take a step. **The only way to get from A to B is to leave A.**

## Is it God's Timing?

I love this one. I do not want to change for fear I will offend the Lord. The Bible is a book inspired by a Spirit in motion, written by men in motion, about a God who is in motion, to a people who are expected to be in motion. When we open our hearts to change, the Word becomes a light unto our path. It is not a lamp unto our parking lot. The Lord has given His Word to guide us as we go.

Jesus said, "No longer do I call you servants, for a servant does not know what his master is doing; ... but I chose you and appointed you that you should go and bear fruit, and that your fruit should remain..." (John 15:15,16). These verses tell us several things about the going process and why we should be comfortable in it.

**First, it is based on friendship.** Jesus has included you in a "Life Circle." Your walk with Him is relational. He does not relate to you as a slave, having no honor. He relates to you as a friend. He will return your call.

**Second, you are in on the big picture.** That is to say that your going is directly related to His coming. Steps in God bring us nearer to the day He will come. You are walking away from a place in Him to a new place in Him. In Him are many abiding places, but they are all in Him.

**Third, He has gone before and is calling you to Himself.** He selected you for this purpose. He has placed you in the "Life Circle" and is moving the whole group forward in His purpose. You were not alone, you are not alone, and you will not be alone. Lisa Greene has a "Life Circle." Pastor Greene has a "Life Circle" if he wants it. Only Pastor Jon does not because he chose the pyramid.

**Fourth, Jesus has ordained your steps.** "The steps of a good man are ordered by the Lord, and He delights in his way. Though he fall, he shall not be utterly cast down; for the Lord upholds him with His hand" (Psalms 37:23,24). "And we know that all things work together for good to those who love God, to those who are the called according to His purpose" (Romans 8:28).

You do not have to be your own guide. You do not have to manage your next step. You are not doing it for Him; He is doing this for you.

**Fifth, the purpose of your relationship is to get you into motion.** You were stagnating in sin. In fact you were dead in sin. You were so stuck in one rut that they had covered you over with dirt. The filth of the world had filled your nostrils and you were suffocated in sin. Jesus, on the way out of His own tomb, brought you out of yours. He has transferred His motion to you so you are now headed to the same destination. There is a place appointed of the Father for you and you are just sojourning in the earth until you get there. When your permanent place is in Christ, you are just passing through all of this life. You are in His perfect will and timing because He has ordered the path and pace of your feet.

**Sixth, it is going to work out.** You will bring forth fruit. Your life will not be wasted. Your fruit will remain. You can have confidence in Christ to direct your path. He has appointed good things for you. He will cause you to prosper. When your life is wrapped up in Jesus, He promises a life of purpose, a life of praise, a life of productivity.

As Jesus prepared to leave, the disciples asked Him, "Will you now restore the Kingdom?" He explained that it is in the hands of the Father and no one knows the times or seasons. You will never know. I personally think that we are about 2,000 years behind the perfect plan of God. It has been the spirit of man saying, "Stay" that has caused us to be so far behind.

## Loneliness

**"I do not want to be alone."** Loneliness is a horrible thing. When Megan was born, Robert Greene had felt alone. Here was a change for which he was not at all prepared. His life had been sailing along. The initial debt of the church had been cleared; the board had approved a nice salary and expense package for him. He and Lisa were very excited about this unexpected child. James and Jainine were model children. His world was all set.

Then came Megan. The doctor had said that the problem was fibrous cysts in her lungs. They had to be broken free so that she could breathe. It was life threatening. Robert could not cope with this sudden change. He was used to a give-and-take relationship with his other kids. He got as much life from them as he gave them. It was fair. It was fun. They were always together. He loved to see the van pull into the parking lot and the two tumble out and come racing to his office. Mary, his secretary always kept the calendar clear for this joyful interruption.

Once Megan arrived, they got the second car. Lisa had to be at home with her always. The kids joined a car pool so that he would not have to stop his day at 2:30. There was no longer a happy greeting for him at home. Instead, a totally stressed wife and two very anxious little ones were his greeting party when he dragged in at six.

Lisa blamed herself for Megan's condition. "Why had she wanted a third child?" was her constant question. "Weren't the first two enough?" She had not been able to clear her mind of the guilt she felt. As a result, she no longer related to him as she had. He felt abandoned. He felt guilty. He was angry with God about this. He could not make the change from Perfect Pastor to Suffering Servant.

The first time he had driven through the red light district was on a whim. He had left the office at nine after a board meeting and instead of going home, took his loneliness to the street. Luke and Gene had asked if he wanted to stop for coffee on the way home, but he wanted to be with people he did not know. He stopped first at a bar, but the bartender called him by name when he walked in. He had ministered to the man's son a few years before and the man announced to everyone at hand that this was Pastor Robert Greene.

As he had left the bar, he had noticed a woman about his age walking along the street. He had offered her a ride and as they went along, had begun to share with her the loneliness he felt. She too was lonely and they began to console each other about the changes in their lives. Thinking they had reached her home, he followed her suggestion to park the car. Enjoying the company, he had thought little of the way she remained in the car.

They shared intimate feelings and, without restraint, fell into each other's arms. At first it was innocent to him, but the surge of desire carried him into sin. When she did finally leave the car, his loneliness was married to guilt and pregnant with despair. He sat for a long time gripping the steering wheel and staring down the road. It had been the patrol car's lights that had broken his requiem. He was dead.

Robert had been defeated by loneliness. He had tried to find fellowship in a world so foreign that he had fallen prey to its oldest sin. He had not been able to let go of the way he thought life should treat him and grasp the reality of what his life had become. Desperation had driven him to hell's door. He was among the dogs of the earth and, before he knew it, he was deeply in trouble. It had happened so very quickly. He had not any sense of emotion toward the woman. He didn't blame her. He had made the drive. He had declined the invitation to stop for coffee. He had turned to the world. It had been his choice.

Robert Greene's greatest sin was turning from Christ in a time of loneliness. Jesus knows that change is difficult for us. After all it was He who left heaven to come to earth and redeem us. It was He who had cried out, "My God, my God why have you forsaken me." It was He who agonized alone in the garden as his close friends slept just a few meters away. It was He who had no relationship with a woman. It was He who was used by us all on His way to the cross. Jesus knew more about loneliness than anyone who ever lived.

When He commissioned the Apostles to a life of constant change, He told them, "You go into all the earth … and I will be with you." He had said to us all, "I will send you another comforter." He does not leave us alone. In the depths of loneliness we must cry out to Christ alone. He it is who will save us. He it is who will keep us.

Robert Greene would now face the biggest change in his life. He would go from the top of the pyramid and free fall as far as his pride would take him. Loneliness can drive us to or from Christ. Robert ran away from Jesus until he tripped and fell.

*Chapter 15*

# Struggles of Going

ROBERT'S dilemma is seen also in the descendants of Noah. As we find them in Genesis 11, they had been spared by the grace of God. They had been given plenty to meet their needs. They were in one mind and language. Literally, "One Lip."

In accordance with the will of God, they were traveling and filling all the earth. They went toward Shinar, and they began to live there. Evidently this was a place of provisions sufficient to sustain them for some considerable length of time. They settled in and made more permanent dwellings of brick, and stone. In a word, they stopped.

They built houses with brick and mortar which was a slimy sort of adhesive substance composed of dirt, straw, and dung. This form of construction was common to the post-diluvial era.

Their spirit of man was then revealed in the suggestion that they build a tower. This also was common to the time as each stronghold or emerging city was signified by a tower to which they could run in case of floods or attack from other people. They were, however, unified, so one has to assume that they had a purpose other than the practical.

Genesis 11:4 tells us clearly what was on their minds, "Come, let us build ourselves a city, and a tower, whose top is in the heavens; let us make a name for ourselves, lest we be scattered abroad over the face of the whole earth."

**There are six more issues you will have to resolve to in that little word, "GO."**

## Gathering in One Place

The smallest word in our acrostic may just be the most difficult. The desire in man to stay in one place is so very strong that only your word, worship, and work disciplines will overcome it. And, the greatest work you may have to do to move your group from A to B is to get yourself going. Objects at rest tend to remain at rest; only objects in motion tend to remain in motion.

**"Come"** — people love to live in piles. It would appear that over all the earth the same thing is happening. People are moving from the country to the city. This has resulted in the shanty-towns of Latin America. It has caused cities like Bangkok, Mexico, Los Angeles (LA), Toronto, and Atlanta to become sprawling centers of corruption. In LA county schools, 173 languages are spoken as people have gathered from all the earth in the hopes of finding wealth.

Several years ago we studied this trend and found that in Guatemala, it took just six weeks for the average immigrant from the country to become involved in some illegal activity. The border patrols in the US nightly chase people back from the cities. Man wants to settle in a place of perceived abundance and get his share of the wealth. It is in his nature to do so.

Certainly they are in motion, but not a positive purpose motion, for, like Robert Greene, they are running away from the direction of the Lord and to their own desires. They are calling to one another to provide their own protection.

## Building a City

**"Let us build ourselves a city."** They had an order in mind. They wanted to stop the sojourn. Was this what had caused Pastor Jon to bypass the elders of his own church? Was this the reason that, without their knowledge he approached a bank to borrow the money to expand his already multi-building facility? Was he building a city and calling the others to join him?

Pastor Greene could find a person his age and lonely in the city. They are everywhere. Families are ruined. Morals are gone. People are desperate for money and will sell themselves to get it. The city is a tough place. And a city whose builder is man is an especially tough place because it is going to be the manifestation of the lust of the flesh, the lust of the eyes, and the pride of life. That is all you are going to find in most of the world's cities.

One day there will be a city whose builder and maker is God. Those who live by faith are also seeking a city, a better one. They sojourn in the land in this life, not at all drawn to the cities of man. But, when life takes a sudden turn, they have a choice to make.

As a missionary, I have drawn close to the Christian leaders of the countries I serve. We have become close enough that they can ask questions, which usually offend Caucasians. One of the most often-asked questions is "Why do missionaries come and stay in the big cities instead of going out to the villages?" "Go" has degrees doesn't it? "Go" has in it, "How far?" and "For how long?"

We encourage people to go as far as they can. Get as far into the people as you possibly can. Be the only white face in the picture.

Immerse yourself in the language. Get into the country as fast as you can. In the first year, learn the language and make four friends. Travel with those friends as much as possible. Learn their ways. Learn the culture. Let your language difficulties make you vulnerable to the local population and allow the humility gained to make a way for you among them.

## Building a Tower

**"Let us build a tower for ourselves."** This amounts to a declaration of two purposes. One is the visible display of strength; the other, the establishment of stability. Both are going to hinder you from going on in God.

Self-reliance exalts the human spirit in the face of God. It is on this battleground that the impoverished of the world languish. They really do not care who is in power. The promise of a strong man to lead them will drive them to the streets in hope and leave them used and dash them into despair. The inevitable conflicts between two cities destroy lives and scar generations. A study of each culture will show that throughout the history of the human race, war has raged between cities and men have exalted in the size of their towers.

Even in the "advanced culture" of the United States, what is the national symbol if not the Washington Monument? Built by Masons all over the world, this symbol appears in every capitol and on most courts. How can the tower be so universal? Because the spirit of man is the same no matter what color that man may be. The tower suggests that, "As long as this symbol stands, we are permanent and in control."

Throughout Europe the founding great father of cities is mounted upon a horse on the top of the tower. In Asia, temples are built around the city tower. In Hindu lands, the "lingum" is raised up to bring contact with the heavens and once a year the towers are anointed in symbolism of that union. This symbol, more than any other represents the common desire of man to declare himself in the earth.

## Reaching to the Heavens

**"Let us build it right up into the heavens."** This is going to be difficult for your "Life Circle" to balance in order that you might go on in the Lord. You are going to be in touch with God. You are, as you pray together, going to reach to the heavens. That is a good thing. But, you must not build a tabernacle around the experience or you will, like those before you never become the global impacting ministry God wants you to be.

Pastor Greene built his tower, but it could not protect him. Pastor Jon wanted to build his tower, but his elders withstood him. Pastor Dowdy has a magnificent five-story building and 4,000 members, but she is moving on. Indeed, they hear from God in that place, but it is not their end point, they have a progressive vision.

## Making a Name for Ourselves

**"Let us make a name for ourselves."** This, the fifth issue is one of the things with which Robert Greene and his new "Life Circle" are going to have to deal with. The thought that several hundred people drove past many churches to hear him speak gave him a sense of invincibility. They even drove past Asa and Pastor George's churches on their way.

Pastor Robert Greene was so well known that the police and magistrate delayed taking action long enough to give him warning. He was so well known that the bartender identified him as soon as he came in the door. His ministry had given him a great name in the context of his community. That name was about to come down to the glee of those who hated his message. The fall of it was about to be thrown in the faces of so many who wanted desperately to see their loved ones come to Christ. In some circles, that name would forever be a source of scorn.

Instead of moving on in the Lord, Robert Greene and Pastor Jon had decided to make names for themselves. It is natural to do this. It is a part of the spirit of man. Each nation, each business, each leader, each political party, makes it a priority to make a name for themselves. It is the way of the world.

Going into a new place where you are not recognized is very intimidating. So often as I change nations and cities, the "Who do you know?" game is first played. I understand that people want a point of reference. I have had the joy of meeting many wonderful people, but I feel that I "know" very few. I know Jesus, and He knows me, but man wants reference from man.

Robert Greene will think about his name quite a bit as it flashes across TV screens and in local papers. Lisa and the children will also have many opportunities to think about that name. He had made a name for himself, but now it was going to be changed. There would be times when he would hate the name he had worked so hard to establish. The church would have to go to a new name; its instead of his. If he left it, it would suffer through an identity crisis as it learned to be led by the elders.

Pastor Jon had already thrown the church into this transition. At last report they were down to 300 and having difficulty maintaining the budget. The "new man" is a gentle, broken spirit who would not think to have the place identified as "his." It will be interesting to see for how long. What will arise from the ashes, a new branch or a new pyramid?

## Avoiding Scattering

**"Lest we scatter over the face of the whole earth."** This sixth issue is the statement of the rebellion of human nature to the demands of God. He said to go into all the earth. They had failed to do so and He had sent the flood. These survivors of the flood now manifest the same nature as their parents. They do not want to go into all the earth as the Lord had directed. They are determined to remain in one place and make a name for themselves. The spirit of man has dominated the spirit of God and they are doomed.

It is this strong human spirit that causes people to stagnate in their lives. They become comfortable and complacent. Their present situation, no matter how limited, seems safe and, in their microcosmic existence, they have made a name for themselves. In the context of Christ, this is a disaster.

The Lord's work is expanded through the relocation of people. The BRIDGE you are building is vital to that expansion. He is very clear that we are to go into all the earth and make disciples of every nation. Yes, to be successful in this you must have a personal directive, but that directive comes to the heart that is set to serve Him. The spirit of man rules the heart that is content to stay in one place and make a name for itself. Its fruit is that of the spirit which dominates it. It has been, since Adam, the desire of the Lord to fill the earth with His glory. Pastor Jon missed the opportunity to move into the six towns in which his elders lived because he had to control everything. The lost had to come to his tower for provision. Now it is no longer his tower.

## Enlarging Your Tent

Pastor Dowdy has established the **Global Business Network,** which delivers multifaceted leadership development into many nations. She has taken the spiritual food which was accumulated in the tower and is sending it all around the world. As she says, "If a person tells me they just want to remain as they are, they are saying that they just want the Kingdom of God to remain as it is." They have a small vision of the Kingdom of God. But if they are willing to go on in Him, they can be used to expand the Kingdom."

One of Robert Greene's issues was his complacency with the size of the work he led. He was "the man." Everyone came to him for wisdom and to know the will of God. He controlled spiritual, human, and material resource. Although his kingdom was established, it was very small compared to the vision God had for him.

## Let God Show You the Way

As you and your "Life Circle" respond to the call of God to go, you are going to face these six issues, **"Gathering in one place"**, **"Building a City"**, **"Building a Tower"**, **"Reaching to the Heavens"**, **"Making a Name for Ourselves"**, and **"Avoiding Scattering."** How you handle them will determine the strength of your BRIDGE. Let me encourage you with this thought — Jesus left his throne and came to earth, He fulfilled His purpose and returned to the place from which He had come. He is the one who will keep you as you obey Him. He knows the way. He will show you the way. He is the way. Walk in Him and He will keep you.

God had an answer for those in Genesis 11. He came down and visited them. He tore down their tower and confused their languages so that they had to go into all the earth. They could no longer walk together. In the confusion of the tongues, He ensured that they would not be building any towers.

As a result, they were spread abroad over all the earth. This was His plan from the start. He had to get involved to push them to where they would not otherwise have gone. He is the agent behind the instrument.

God is also moving Robert Greene. During the long and lonely nights Lisa and Megan agreed in prayer that Robert's sin would be uncovered. Through the months of waiting for God to answer, Lisa developed a poise that was born of the Holy Spirit. In her times of questioning, the Lord met her. It was this poised preparation for change that Robert saw as she stood in the doorway. The Lord had given her grace to change, no matter what the cost.

The board of the church also had the grace to change. Luke almost left before they had prayed through, but his wife kept him there. They are going to have to minister in that grace if they are going to walk across the bridge with Robert. He will never be able to go back to the way things were. But, together, they will be able to walk on to a new place in Christ.

Pastor Jon's elders are not willing to walk the bridge. The lack of relationship with him has cost the future of the work. He will walk alone and they will decide again about walking together. Perhaps this is the time they will each build a "Life Circle" and seek the Lord about the way He wants them to go.

You also have choices to make. Of course there are issues, but do you want to go on in the Lord? He certainly has not stopped calling to you to follow Him. He still wants to make you into a Fisher of Men. He will never stop calling until He has called you into His eternal presence. So, you must go on in Him. Face the issues, form a "Life Circle," seek the Lord together, do what He tells you to do. His grace will be sufficient for change.

*Chapter 16*

# Leading From Within

"LEADERSHIP is the discipline of exerting special influence within the group to move the group toward goals of beneficial permanence that meet the group's real needs." This definition by John Haggai as found in his book *Lead On* has become the standard for our lives. It has caused the reconfiguration of our understanding on how successful ministries and businesses operate. It has helped us to move within the Body of Christ without causing division and disappointment.

## Benefiting the Group, Not the Leader

Leadership exerts influence to benefit the group. Manipulation exerts the same influence to benefit the leader. The pyramid benefits the leader. The "Life Circle" benefits the group. The influence exerted is twofold. First, there is the influence that God gives. Second, there is the influence given by the group.

In the "Life Circle," the leader sets the agenda, leads the meeting, teaches or shares from the Bible and encourages the input of others. The Leader is responsible for keeping the others on track and to be certain that all have an equal opportunity to participate. While the Leader is not the "scribe" of the group, he has the role of verification and communication of those things brought forth in the meeting. This style of leadership will result in the energy of the leader being communicated within the group. It allows for the fruit of the three disciplines to develop.

In the pyramid, each member is given a task, but does not relate to the others enough to build an interest or relational base. If a person is perceived as gathering others, the leader will issue a "loyalty statement" of some kind. From where does this insecurity come? It comes from the need to control. The leader in a pyramid must know at every moment that those under his manipulation will not grow strong enough to split away and do their own thing. So much energy is used to control that there is no longer energy to accomplish vision.

In a volunteer organization, those who follow can sense the motives of those who lead. The subordinates will remain loyal for a season, but unless they are given a clear statement of goals, they will eventually lose heart in the vision of the organization. This is what has happened with Pastor Jon and his elders. Trust was broken through a selfish act. It was not the first time. It was a pattern with which they had become familiar. When I asked them to repair the joint that was out of place, they replied that the tear in the relationship had become too great for them. They just could not rally again to the vision for their trust in the leader had been completely destroyed. When he lifted himself above them in self-service, the shadow of the pyramid was cast across the light of their ministry. It is now cast into a season of obscurity and will remain there until, from the ruins, it is lifted once again.

His appeals to "loyalty" are falling on deaf ears. The elders will no longer trust him to exert influence over their lives. The trust that has been violated can be repaired only by a prolonged time in a "Life Circle." They are not going to continue on together, and he is moving on, so years will pass before healing can occur.

## Restoration of Trust

Pastor Greene has a different situation. He has lost his ability to lead. His family is looking to Lisa for leadership. The board members are looking to one another with the help of the two Pastors. The work will stop any forward motion until this trust can be re-established. How long will it take? This depends on his commitment to the process. He will have to become re-involved in the discipleship process. He will have to cut away the compromised areas of his walk with the Lord and develop again his basic disciplines. If he wants to do that, then he will change and the work will limp along until the joint is restored and then run full speed upon it. The joint will heal to the degree to which he desires to submit to discipline. Should he decide that it is too much and he should just leave altogether, studies have shown that it will take five to seven years for the work to gain forward momentum.

Lisa Greene has a big decision to make. If she is going to establish a BRIDGE to span the gap in their relationship, then she is going to have to exert special influence in her "Life Circle," the family, to move them all toward specific goals of beneficial permanence. They are going to have to forgive, and continue to love their father.

For each member of the "Life Circle," there are going to be major decisions. There will be struggles with values. Going on with

Robert will mean a certain embarrassment. Always they will want to cover the truth instead of bringing it into the light of the public eye. Many times they will be asked, "Do you still attend 'that church'?" Bill is correct; the greatest scrutiny is going to come from the members themselves. They are the ones who were used in the pyramid. They are the ones who are smelling, and tasting the death of the organization. Some will run. Others will try to stand but, in either event, trust will have to be relocated in the center of the group.

Christian people are trained to follow. From the top of the pyramid they are taught that they are not mature enough to lead. The Pastor or a few trusted teachers are the anointed ones who will lead the flock. Often the example given is the leadership of Moses and the children in the wilderness. When our little pyramid headed out to Latin America, the church applauded a vision and a song by one of the members. In it, he had seen me as Moses, leading the people to the Promised Land. Affirmed by the Pastor, this became the banner cry for the mission. The only problem was, Moses never made it to the Promised Land, and neither did the people.

Moses hit the rock. The people hit Moses with complaint and criticism every day for 40 years. The people all died under the system without ever getting to the Promised Land. As a former Moses, let me tell you, decide today to get in the middle of a "Life Circle" and find out how wonderful the Promised Land can be. Being Moses is no fun. It also does not bring success.

## The "Going" Process

The departing Admiral of the United States Women's Navy was a great lady. Among other accomplishments, she had been the energy behind the computerization of this, the World's strongest and most advanced fighting force. She is known for having coined the word "Bug" for problems in data transfer. In fact, the problem in that case was an actual bug that had crawled into the case of that primitive processor and caused a short circuit.

She made a very astute observation about resources and administration. **"Material resources can be administered," she observed, "But human resources must be led."**

**"Administration is defined by Webster as, "The application of resource to need."** Doesn't this sound like the view most of us take of leadership? We develop human resource and send it to the need. This is the justification for the "step by step" or "base by base" paradigm used in many churches today. The problem arises when the application goes beyond those four walls. As long as we are just making pew sitters or happy listeners, or participants in practical relationships at the local level, the paradigm is fine but, if we are going to make Fishers of Men, it breaks down.

Jesus said, "Follow me and I will make you..." That means that leaving and going are a part of the process. By Matthew Chapter 10 and Luke Chapter 9, the disciples were the ones going. He had led them through a process. He had replicated the going process in their lives and they kept going with it. This process has four phases.

**First, you go.** That is right. You go to where you want the group to come. If you have a healthy "Life Circle," you can all go. But someone has to go to see where it is you are going to land on the other side of your BRIDGE.

**Second, you go and they go with you.** As you have seen the vision, you form a "They." This will be a "Life Circle" designated to move from where they are to where they are going. They will develop the vision, proclaim the vision, resource the vision, intercede for the vision, prepare the planks for others to reach the vision, go to fulfill the vision and evaluate the vision's success. These are very important people.

**Third, they go and you go with them.** You are now riding on the synergism released through the "Life Circle." They can carry it, because you taught them from within. They will sustain the work. If you have exerted influence within the group, they will be ready for this transition. How I love to visit those who are planted among the nations and see their prosperity in the Lord's work.

**Fourth, they go on and you do too.** They build BRIDGES to the lost around them and you see a new place and begin again. And so it goes from place to place until all the earth has heard the good news that Jesus is risen from the dead, proving that He is the One whom we can trust.

## The Relationship Factor

About the same time as the retirement of the Admiral a survey was done of all commands in the US Navy. This would mean any place where personnel were engaged in a specific mission. It could be as small as a dispensary or warehouse, as expensive as a jet warplane, or as vast as an aircraft carrier with 5,000 on board. The "Successful" command was that which, "Reached its goals in the time allowed, within the available budget." Imagine if we evaluated our work by the same standard.

The Navy found one key to successful commands. It was not the size of the budget. It was not the number of personnel. **The Navy found that the success of the Command was determined by the relationship between the commanding**

**and executive officers.** Where the commander "lorded" his position over the subordinate, the command did not succeed to the same degree as those in which the Commander became a servant to the Executive Officer and brought him to an equal professional level. Of course, the final decision was with the ranking officer, but those commands never came to a "final decision" crisis, they navigated securely through the seas of change, accomplishing tasks, developing personnel, leading the free world. How important is a "Life Circle?"

Robert Greene did not give place to the "Executive" in the church. He belittled the men with whom he was to serve the vision.

He did not give place to the "Executive" in his home. He was unfaithful to his wife and children.

He did not give place to his professional colleagues, thinking of them as "losers."

He did not accept the authority of the Chief of Police or the District Magistrate when they tried to warn him of the consequence of his ways.

Robert's pyramid had become so grand that the only voice he could hear was his own. Even the voice of his soul was deafened by the clamor of his self-pity.

Robert had fallen into the trap that befell King David. 2 Samuel 11 tells us that in the spring of the year when Kings go out to war, David remained on his housetop. From there he saw a woman bathing. He wanted her, and arranged to have her. He sent into battle those men with whom he had hidden in the caves. He sent into battle those who had formed a literal "Life Circle" around him to defeat the Philistines. He sent into battle those men from whom he felt unworthy to accept a cold drink because he was humbled at their commitment to him. Those men he sent into battle while he rested upon his housetop.

## Serving in the Midst of Your Flock

You have no idea how many missionaries I have visited who have been deeply hurt because the pastor with whom they served so faithfully has never taken the interest to visit them on the battlegrounds. There are literally thousands of them serving around the world who have never seen their Pastor make a visit. When I approach the Pastors, they recoil as if the missionary was being disloyal to them. I feel the shoe is chafing the other foot.

**When a leader says to me, "You must understand, we are now a sending church. I no longer make trips, we are sending younger people,"** I ask, "Who is the Bathsheba?" What desire of the flesh are you serving, that would cause you to leave these people at the mercy of the enemy as you recline in luxury atop your pyramid? Of course they get angry. Of course they are offended, I mean, who am I to ask such questions? But, as long as there are Robert Greenes in the world, I am going to continue to ask.

How did Pastor Jon get his "Let's put the food and clothing at the church and make them come to it" mentality? He did not want to go out to the warfare and take the resource to the need; instead, he wouldn't even send it. He made the broken crawl to him. That sounds more like Pharoah and the famine than Jesus and the 5,000.

Pastor Dowdy has the right track. She has raised up a great "Life Circle" at Trinity Christian Center and now she is exerting her influence to form a new, global, "Life Circle" called the **Global Business Network**. After listening to business and church leaders around the world, and receiving wonderful insight from the Lord, she has partnered with them to form this networking community with global impact. Being able to move from pyramid peak to "Life Circles," she has released new life to thousands who will feel the impact.

Her Executive personnel have access to her. They have seen her strengths and weaknesses. She does not hide, but rather searches out the pastor or business leader who needs the power of the Holy Spirit and a stepladder to climb down from the pyramid upon which they are dying. She carries that ladder to them and carries them in her prayers' helping them to avoid the disaster of denial. She is leading that portion of the body of Christ in which she has found influence into a new day, actually an old day revisited.

Because she realizes there are two types of influence, that given by God and that given by those around us, she can walk in humility, within the group. There is no doubt that her influence is felt, but it is in supportive, not legislative ways. I have had the joy of observing the transition over a few years. They have addressed the issues of "Going" and have been able to form a "Life Circle" of leadership, forming the BRIDGE over which the Global Business Network will travel to their Promised Land of the 21st Century.

*Chapter 17*

# Evaluation

ROBERT Greene is about to have a finger in his eye. He didn't want to have it happen. He had become an expert at keeping people away from such sensitive areas. Even Lisa had not been able to probe into the recesses of his life. Her attempt to bring him back to the Word Discipline had resulted in acid sarcasm. There was no way he was going to allow her to become his judge. He drove her away with bitter words and cutting looks. Without a "Life Circle," he had no way to have his eyes washed out. His vision was lost, his personal ministry a disaster. His life was a birds-nest of unfulfilled desire. From his pyramid perch he could no longer see his way.

**Successful ministries are those who reach their goals in the time allowed and within the available budget.** Robert and many like him have lost their way because they do not have a "Life Circle" defining the progressive vision of their organization. They have not defined their product, so they cannot define the process.

**The E in our Bridge is "Evaluation."** In these chapters we will talk about SMART Goals and the application of the Base,

Resource, Intercession, Disciplines, and Going process. By the conclusion, we will know how you are going to build your Bridge to get from A to B.

## Evaluating Success

**Jesus said, "Follow Me and I will make you Fishers of Men."** He defined the product at the time of recruitment. The process was going to take three years. At the end of that time His "Life Circle" would be able to sustain itself and accomplish the purpose for which He had come, to tell the world of His love for them.

Christ's success is seen clearly in His report to the Father. Could there have ever been a closer commander-executive relationship? In John 17 we can see the details of His report.

In verse one of the report, **Jesus refers to the temporal aspect of His work.** Like Pastors Greene, Jon, and Dowdy, His task is to communicate eternal life principles to a temporal community. Spiritual truth is timeless. Temporal norms change as cultures develop. The person who is going to succeed in Bridging this difference is one who, like Christ, in verse two, has been given all authority over the flesh. The Greek for "authority" here is, literally, "freedom". The sustaining power of the "Life Circle" is a way to keep that freedom.

**Jesus then expresses His realization that His part is in the context of the whole.** His picture is huge. He is the only one in the history of the world who could do what He has done. He is the Champion of our salvation. But, He phrases His evaluation in humble terms, "The ones you have given me." His success is in terms of His relationship to the Father. He does not see Himself as an independent free agent. He keeps His worship discipline and humility by declaring that His success is because of the Father giving to Him. This humility will hold Him to the cross as He despises the shame, looking forward to the joy that is set before Him.

## Setting the Vision and Mission

He then gives the simple statement of their vision, **"That He should give eternal life to as many as you gave Him."** There it is. He has become the BRIDGE to eternity. Life has passed from the Trinity to the Disciples over the BRIDGE who is Christ. He said to Nathaniel, "You shall see the angels ascending and descending upon the Son of Man." He is the ladder in Jacob's vision. He is the Bridge over the chasm of hell. He is the One who embodies the principles you have learned. It is Christ who has become the Way to the Father, for no man comes to the Father, but by Christ.

Jesus restates the definition of His mission in verse three. **"And this is eternal life, that they may know you, the only true God, and Jesus Christ whom you have sent."** How did they get to know the true God, and His Christ? By being part of a "Life Circle." Jesus did not stay apart from the disciples. There was not professional distance or safety zone for Him. He came and walked with them, ate with them, talked with them, slept where they did. Jesus identified with those in His "Life Circle." There was no evening board meeting and going home in separate cars for Christ. Life was a perpetual fellowship. He knew their questions and struggles, and He allowed them their weaknesses. His ministry was open. His life was open to them. He came not to be served, but to serve and the service was personal even to the washing of their feet.

## Let God Order Your "Life Circle"

As you form your "Life Circle," do not do it in practical, or ministry-referenced terms. Ask the Lord to join to you those with whom He wants to build in your life. He has a vision. He will add to the church daily those who would be saved. He will build the church upon the rock of Christ.

I love the thrill of prayer walking a new city. I go there and just start to walk and pray. The Lord brings people my way in what we call Divine Appointments. After several weeks there will be a "Life Circle" or Prayer Cell committed to reaching that city with the good news of Jesus. They will be getting to know each other and God. This is the purpose of the ministry of Christ. As we grow in relationship, I introduce them to other cells in other cities and countries and we become a network. We build BRIDGES between cells and over these flows life to the "Life Circles" and to those who live in "their" cities. It is through these relationships that SEAPC has grown into its destiny of "Filling Global Gaps."

In verse four Jesus reports that He has **"Finished the work you have given me to do."** Would any of our three Pastors conclude that the 12 disciples, one of who was an appointed turncoat, were successful in ministry? I mean, for the Son of God to establish 11 people in three years would not be, by today's standards, deemed successful. But, Jesus could declare, "I have finished the task." How did He know?

## Establishing Goals

He knew because from the start He had a specific goal, and a specific timeline to accomplish it. His objective was to redeem man out of every nation. That objective is still being reached today. But the first goal was to establish a "Life Circle" on the earth. That is to say, a relational base upon which He would build His BRIDGE to eternity. When Peter declared Him to be the Christ of God, the relationship was established. As James and John sorted out their roles, the relational nature was developed. As the others were added to the number, the base became multiethnic and multilingual. Jesus' first goal in reaching man was to reach men.

If an individual or ministry does not have a specific goal, the followers will, eventually, lose heart. Because he was out of joint, Pastor Jon lost the confidence of the elders and the people. Resource flows across relational lines through information, inspiration, and invitation. Because the elders could no longer trust him, they could not inform, invite, or inspire the people to give. In fact, they would not be personally involved.

## Specific Goals

**Goals, those small steps to a great march, must first be specific.** Pastor Greene no longer set goals. He just wanted out of the pressure of the pyramid. He was finished; no longer seeing a reason for progressive vision. The pews were full. The bills were paid. The board was peaceful. The church had influence in the community. He had no need to talk of "vision" or "cause." Two in ten of the visitors remained in the church for a season and those who left were just malcontents who could no longer submit to his leadership. "I am not responsible for them," he would say to himself. "If they want to go, I am not going to do anything to try to keep them here."

Now Pastor Asa will be setting specific goals for Robert and it is just as well that he will be. At least 25 percent of the "malcontents" are now in "Life Circles" served by Asa's pastoral team. They have very specific goals, meeting weekly to pray for one another and to hear the emerging vision for their "Life Circles." Soon, should he submit to the discipline, Robert would find the joy of taking small steps of righteousness that will bring him back to the light.

**Smart goals are specific goals.** Ambiguity flees when specific goals are announced. For your Bridge to be effective, your "Life Circle" will have to agree upon goals that will move the group to beneficial permanence that meets their real needs.

## Measurable Goals

**Smart goals can be measured.** Naomi Dowdy pastors a church in Singapore. There are larger churches. There are much smaller churches. There are churches with larger facilities and with more emphasis on areas of ministry to which Trinity has not been called. Naomi has taught her "Life Circle" to measure effectiveness against goals, not against other ministries. The measure is the accomplishment of specific goals in the time allowed and within the budget.

Paul writes to the church at Corinth, "You compare yourselves with yourselves, which is unwise." We say, "Comparison is the basis of condemnation for in every comparison someone is lifted up and someone is put down." We avoid comparison against other ministries and ministers. We declare what the Lord has given as our objective and we state specific goals that will get us there.

Robert Greene compared himself to the other Pastors in his town and thought of them as losers. He had no respect for them.

In his need to justify his own actions, he had to remove the standard of comparison. It proved to be a disaster. Moses had no peer. Yet, in his humility, he referred to himself as the meekest man on the face of the earth.

If the meekest man could hit the rock, what possible chance does any other "Pyramid Percher" have? It is going to happen. If you do not have a fellowship of peers with whom you can be open and who have the ability to work with you for specific goals, then you are going to hit that rock for sure. You cannot avoid it. It is going to happen and when it does, you are going to be as miserable as Robert Greene is.

## Attainable Goals

**To be a smart goal, your goal must be attainable.** Pastor Jon's goals could not be attained with available resource. The elders tried to communicate that to him, but they were unable to bring understanding. He then began to think of them as lacking faith. They thought of him as lacking humility. All they were asking was time to develop trust to a new level so that together they could extend their faith for the goal or next step to their agreed objective of reaching their surrounding communities. Pastor Jon could not give them the time. He made a quick move and will regret it for years to come.

Hebrews 6:12 tells us that through faith and patience we shall inherit the promises of God. This great combination is the heart of waiting for attainable goals. Some goals are just too ambitious for today. The "Life Circle" needs to say, "Maybe we can be ready for that step in four months, but for now, we are not ready." Leadership has to realize that we all go in together.

## Do Not Overextend Your Resources

Chinese leadership style teaches that we never commit more than 75 percent of resources toward a goal at any one time. So many problems are caused by overextending spiritual (we just haven't prayed enough), human (I hate to say no, but I cannot squeeze one more meeting in each week), and material (all they ever talk about is money) resources. Wise is the leader who understands how to build the "Life Circle," praying in enough core resource to establish attainable goals.

Let's say you are helping a missionary raise an annual budget of US$50,000. No, he does not need that much to live, but the abundance would release him to new areas of ministry. In the Assemblies of God, the average contribution to missions is $80 monthly. If our man can gain one new donor weekly, he will reach his annual goal in one year's itinerating. And, as a bonus, he can establish 50 new prayer cells, a great gain on the Spiritual resource side. If he works through local churches and networks, he may be able to gain 30 or 40 people who will come and visit him over the next term.

Approached in goal-size slices, this seemingly lofty vision can be attained without a great deal of exhausting work. He will have to keep his disciplines in order, but that should not be difficult if he realizes that they are building the BRIDGE over which thousands can come to Christ. To a great measure his success will be determined by his attitude toward those around him. Should he see himself as exerting influence within the donor group to move them to benefit, he will be very successful.

Does he believe that those who participate financially in mission will share the reward of the harvest? Does he believe that he is dependent on the hard work of others for his daily bread?

Does he have a grudge against those who do not join him in the work or can he see that there are differing gifts in the body? Does he resent the "security" of a local pastor or is he thankful that he does not have to serve in this way? In short, does he radiate joy or bitterness? If joy, then he will be very successful for the Lord loves a cheerful giver.

## Realistic Goals

**To be a smart goal, your goal must be realistic.** How do we determine what is realistic? Can it be reached in the time allowed? Can it be reached within available budget? When I wrote our first book, *Every Place Your Foot Shall Tread*, I did not realize several things. First, it takes longer than I thought to run the book through an editor or ghost-writer. Second, it takes much longer to sell 10,000 copies of a book than I ever realized. Third, you make much less money on a book than I ever realized.

I made the mistake of accepting the offer of friendly, interest-free loans to finance the completion of the book. I thought that the book would be finished quickly and sell quickly and that I would realize sufficient income to pay the loans and underwrite some of the work we are doing in Asia. I made a mental commitment to both.

Thank God for faithful friends who are still faithful friends even though it has taken much longer to repay them. If I were not in the center of a "Life Circle," I would have gone under with the pressure of it all. While relationships were pressured, by clear communication they were preserved. It gave me the opportunity to be humbled rather than lifted up by the project, but it was definitely mismanaged. The fault was mine. I was not realistic in my time line and cash needs.

Where do faith and good planning come together? They combine in the "Life Circle" for there is safety in the multitude of counsel. Wisdom is found in the multitude of counsel. War is waged in the multitude of counsel. This book has passed through many hands before it got to this finished stage. The manuscript has been passed through no fewer than eight sets of hands before it went to editors. It has been checked for Biblical as well as practical accuracy. And no area of ministry was waiting for the money you paid for this book. This book has been screened, checked and cross-checked. The money to publish and print was prayed in. Yes, I wrote it, but the "Life Circle" for books produced it.

The most painful area of ministry for me is seeing needs that I do not have the resources to meet. Starving children are number one, abused women another, and the general condition of those masses who are on their way to hell the third. Not having set realistic goals on our first book created an anxiety in my life that spilled over on others. Desperate people say and do desperate things. People become desperate when they do not have SMART goals.

Faced with the situation in his home, Robert Greene could not set a realistic end to the torture his family faced. The relationships were all redefined by Megan's illness. Doctors could not give them a definite time line for her health improving. Often he and Lisa would leave the office in complete frustration for even though he would have loved to say she would be healed, the doctor could not.

The drives home were terrible. He was angry with God over the condition of his child. His theology did not allow for bad things to happen to good people. Didn't God know that Robert had to be the living example to the church? How was it his deepest prayers were falling on deaf ears? He hit that Rock many times a day. The locomotive of anger pulled a train of frustration whose cars spilled disappointment and bitterness at every turn or bump in the road. His expectations were unrealistic.

"She will be healed by her first birthday," he had thought one day in prayer. Her birthday celebration was punctuated by a trip to the emergency room, as she could not breathe.

"She will be healed when the evangelist prays for her." He had avoided her in the healing line. Evidently his faith was, like theirs, not up to it.

Monthly they danced around the issue with the doctors. Lisa had taken her alone to the past several visits. He just could not do it again without losing his temper. Everyone in his world knew how to avoid Robert's losing his temper. His nerves were frayed and temper at just below boiling. He lived as an active volcano. He could not understand how Lisa had so much poise.

Lisa did not live in the pressure of time lines to healing. She created no false hope; she entertained no dream of a miracle. To her, Megan was a midnight prayer partner. A wonderful child who gave as much comfort as she required. When the fibrosis was first diagnosed, she felt guilty for bringing the child into the world. As she rocked this child, she would ask the Lord what she had done to deserve this. After a month or so she realized during a particularly difficult night that she was not the one suffering the disease. This baby was doing all she could to live.

Lisa set goals. Four hours sleep. A nap when Megan slept after her afternoon back pounding. Forming a "Life Circle" with the other kids so that together they could face this warfare in their home. How they had responded. It was James' opportunity to be a big brother and he loved every minute of it. Jainine realized after a few months that she would not break the baby and began to carry Megan everywhere.

Lisa determined that she would not complain against God in her prayers. In fact, she learned to worship Him in the darkest hours of the nights. If only, she had thought, Robert could speak to or pray with someone who could help him find release from his unrealistic goals. Each time he would announce "The Lord had said" she prepared to explain to the children that Daddy was under a lot of pressure to perform a miracle and finding that he could not do so was very difficult for him.

She had thought of calling the elders, but he had said no. She thought perhaps another Pastor, someone with whom they had gone to school. But he had said no. She suggested Asa and it took three days for her husband to calm down. She wondered at the deep bitterness, but did not respond. His frequent outbursts in their home caused great pressure for the kids, so she just kept still.

Robert had lost the ability to make realistic decisions. His worship discipline was a disaster. His word discipline had failed. He stayed long hours at work. She didn't know if he was avoiding her, the children, or the Lord, but he had no "Life Circle."

## Tangible Goals

**For a goal to be smart, it must be tangible.** Often ministries drift to the "intangibles" for a sense of success. Robert Greene did not any longer have tangible goals. He could not measure his progress because, from the top of the pyramid he was lord of his home, lord of his office, and in the recesses of his mind, lord of his own existence. When given leadership materials that encouraged good goals, he set the books aside with a smirk. When invited to attend an area meeting with a widely-acclaimed teacher on leadership, he smugly assured the host that he already practiced all that was included in the seminar. He had no need of a teacher.

His actual frustration was, until his personal problems were worked out, and his disciplines re-established, he was not going to be able to build a bridge for anyone to cross. In fact, the idea of picking up a new plank was for him most offensive. In having to close one eye in pain, he had lost depth perception. He had given his life over to "intangibles." He was following his emotions down the driveway of despair only to park them in a space reserved for him by the agents of hell.

He was going to have to open that eye to some trusted friend who could remove the splinter. Someone with clean hands and a pure heart, which could cause the tears of repentance to flow. Who could direct that bitter herb of Megan's illness to become a purging hyssop that would restore snow-white purity to Robert's heart?

*Chapter 18*

# Look Again

WHO would you want to stick their finger in your eye and remove a splinter? You certainly would want a peer; someone who understood the pressures you faced and saw your big picture. You would want a person of compassion who would not fault you in the process, but who would commiserate with your malady. You would want a person of integrity who would keep anything you might say in confidence. You would want a person with clean hands so that further infection would be prevented.

The only perfect men are single. All other men have someone who can see their faults, their wife. Robert Greene was not able to allow Lisa to pull the splinter from his eye. "Why," you ask. Because in his heart he wanted out, not of the marriage, out of the responsibility. He did not have the personal disciplines necessary to lead his family through the "going" process to a new relationship, which included the very real possibility of a tragic death in their home. The fear of death had made him subject to bondage all his life. He could not shake it, and he could not approach anyone for, if he did, he would have to come down from the top of the pyramid. It was too great a price.

## Being Transparent Before the Lord

Naomi Dowdy is single. Yes, a very successful Pastor and missionary who is single. She has surrounded herself with "Life Circles" who have the liberty to speak into her life when they feel a need. She is so open; she does not have to be defensive. She will share a prayer need with those who are proven prayer warriors. She will share a personal concern with very trusted confidantes. **Those around her grow rapidly in the radiance of her personality and learn quickly that if you have to hide "it," "it" is probably not a good thing to do.**

How can she be so free to exert influence within a group? She has nothing to hide. She lives in a city given to gossip. She has surrounded herself with "Life Circles." She listens more than she speaks and, when she is the spokesperson for an idea, each member of the "Life Circle" can hear his or her input included in her presentation.

She is very specific. She presents the standard of measure before the idea. She explains how goals are attainable and is exact in her resource evaluations. She is realistic, developing spiritual resources before human and material. She realizes that except the Lord builds the house, her labor is in vain. Each goal she establishes for the network she influences can be measured in numbers of people, dollar amounts, decision cards, and cells established. There is no intangible factor in her goals.

But, there are times when her vision can become obscured. How does she deal with it? She goes to the Lord first. Her worship discipline allows tears of adoration to wash her vision. She goes to the Word for fresh light and insight. She goes to work, exploring other points of view. She studies, listens, and asks questions in the attempt to clear her vision. And, she has a

"Life Circle" of trusted people to whom she can turn to take a splinter from her eye.

When you consider the question, who comes to mind? Is it your spouse? If not, why not? Are there issues that need to be addressed in your marriage or, will you, like Robert Greene, run to others for consolation? Within your ministry or business, are there areas that you cannot see because of dirt that has been blown or thrown into your face?

**Evaluation is the E in our BRIDGE to get from A to B.** There are very few people who enjoy the evaluation process. Many take evaluation very personally. In this chapter we are going to give you seven principles of evaluation that you and your "Life Circle" will need to consider to have success in building your BRIDGE.

## Project Evaluation Is Not Personal

You are not being evaluated. If a "Life Circle" is administering the project, then the synergistic product is being evaluated, not the individuals involved. So often we take the examination of the product as that of the individuals involved and not the process. Robert calls it "My Church" so he is ready to fall into this trap of personalizing the evaluation. When he sees the church as the composite of its members, he will be free from the pressures that cause him to cover his eyes to his wife.

## Project Evaluation Must Be Based on Smart Goals

You are measuring against tangible goals. You have been specific, progress is measurable, it will be clear whether or not your goals are attainable. If goals are achieved early and under budget, perhaps your "Life Circle" will have to raise the standard of the goal. Your goals should cause you to reach to a high standard. If they are too easily reached, they may be as unrealistic as if they were impossible. And, your "Life Circle" must evaluate according to the tangible goals to which you have agreed.

## Evaluation Is Based on Real, Not Emotional Data

Evaluation is not about how you feel or perceive that others feel, it is about whether or not you are meeting your goals in the time allowed within the available resource. While synergism requires the loss of individuality, it produces a release from the pressures associated with the pyramid. Since you are all on one functional team, no individual will gain accolades or condemnation for the results of your team. Authority and responsibility are shared, so evaluation is not personally threatening.

## Evaluation Allows for Goal Alteration

If we find that the climate in an area in which we are working does not allow access for half the year, then we have to redefine our time line. Evaluation will allow for the realignment of expectations, which will ease tension along the joints, which need to function with strength to carry the responsibilities through to obtain the objective. While the objective must not be compromised, the goals that will get you there are open to re-evaluation.

## The Most Experienced People Should Do the Evaluation

In sports discipline, we learn that playing with people who do not intimidate us can give a sense of victory every day, but it is an illusion. Playing with someone whose serve is better, handicap is lower, or average is higher causes us to evaluate the way we do things. When our pyramid crashed, I found three men with whom I could sit and talk. We evaluated my role and the roles of others, as I perceived them to be. In our Thursday morning breakfasts, much time was spent cleaning splinters from eyes so that our perceptions were readjusted. It was really difficult for me to open my heart to these men, but, when I did, the resulting anointing became freedom to as many as 50 pastors and leaders from the area.

## You Cannot Take the Splinter Out Yourself

You need the help of other people. So often I have heard business leaders say to each other, "Let me run this by you and tell me what you think." What are they doing? They are evaluating an idea for probability of success. Those of us who work among unreached peoples fall generally into two groups.

First there are those who are very secretive. They do not share goals or strategies. Therefore their work never undergoes evaluation. They are an entity unto themselves, very territorial, extremely suspecting of the motives of anyone who ventures onto their turf.

The second group are the "networkers." They share information on a "need to know" basis the synergy of their time together is amazing. They form a "Life Circle" for that people group. I have spent time traveling with both groups. As a networker I fit more with the latter, but I in no way mean to demean the work and sacrifice of the first group. Among the networkers, the conversation is consistently, "How are we going to reach that person?" They discuss resources in terms of prayer teams, planting teams, and micro-economic development packages. They strategize from each other's resource pools. They are very synergistic, ready to lose their own ministry identity for the sake of common purpose.

Within the networking community we constantly challenge each other by evaluating plans and results. We call it "road testing". There have been times when a plan has blown up in our faces. Our goals were poorly set and we walked into disasters. Hopefully these have been learning experiences for the entire "Life Circle."

## Judge Nothing Before Its Time

How many projects have been deemed unsuccessful because they looked bleak at some point in their development? I have learned in ministry that it takes 18 months for a positive vision to be accomplished and six months for a disaster. Therefore we encourage "Life Circles" to be setting goals on an 18-month timetable.

We anticipate three negative challenges during the lifespan of any positive project. This provides a regular schedule for evaluation. In SEAPC, we quarterly evaluate each of our projects, affiliate ministries, and those who have submitted for training for ministry. We evaluate based on commonly set goals. The evaluation is impersonal. I know that we are all doing everything we can to reach the goals in the time we have projected. Our evaluation is to determine how we can better serve each other for the accomplishment of those goals.

A report of the evaluation compared to the goals is sent to the chairman of our board of directors, and to major donors. We seek out input from pastors, businessmen, and other foundations. We constantly evaluate, not question, what we do and how we are doing it. As President, I am constantly evaluating our **Relational Base, Resource Development, Intercessory Prayer, Discipleship, and Evaluation processes**. I interview hundreds of people a year to gain every possible eye cleansing solution.

As a server to others in the network, I am challenged by their dedication, and complete selflessness. Trust has formed over time. We are able to daily function without question of each other's commitment to the goals and tasks involved. This has released faith beyond anything we could have imagined.

**Our lives will not be measured by what we could have done with what we might have had. Our lives, businesses, and ministries will be measured by what we did with what we had.** Our success will be evaluated by comparing productivity to resource. We have the best tools and travel any generation has ever enjoyed. We no longer need to bind up resource in buildings, for we can locate at a website and, from there, operate in all the earth. I can speak at a luncheon in Singapore and a dinner the same day in Vancouver or LA. By digital communication and the web, we can be teaching Jesus in every nation all at the same time. Ours is a wonderful day.

Ours is also the generation that will face the greatest evaluation. Medicine and Communications have raised the standard to a new economic pinnacle. Education is the second largest commodity in developing nations. Strategy groups and Core groups have replaced individual leadership. Now we are faced with the greatest challenge, do we have the spiritual and moral fiber to carry these resources to meet the needs?

In this book we have been looking at two people who are faced with the same challenge Jesus carried perfectly. Robert Greene did not have the moral fiber necessary to conduit purity to the community. He lacked the strength to stand in a "Life Circle" and influence it toward smart goals.

The Rev. Dr. Naomi Dowdy has, so far, by God's grace, been able to do so. Now we will see if Robert's newly-formed "Life Circle" can bring him back.

*Chapter 19*

# What About Robert Greene?

THE meeting had started off safely enough. Robert had declined Asa's invitation to ride together and had followed dutifully behind the senior man's Buick. "Funny," he had thought, "I feel like a child being led home from a party."

Several Pastors had gathered in the meeting room of a local hotel. They had come from every denomination in town. Robert recognized most of them, and returned their cautious greetings with equally feigned warmth. It was startling to see how false so many men identified with the "Truth" could be. He felt secure knowing that he could play this game as well or better than they could. There was no sense of intimidation because he was among a group above which he had risen in their vocation. Robert Greene was without peer in this room.

Asa introduced them to one another even though they were well acquainted. He was the master at recognizing all so that they could come to a common ground. By acknowledging each man, he tore down a pyramid. Each man had value in the meeting. Each man was established, not in the numbers of people in his church, but in the fact that he was a follower of the call of God.

They opened with a song and prayer and then the time was passed to Robert to make his confession. Palms strangely damp, he began.

"I am very sorry to have to come to you today to apologize for the sin I have committed and the reproach I have brought on you all and the churches in our community. I ask your forgiveness." Few words delivered in the best contrition he could muster, having no respect for the audience.

Glances flashed across the room. Who did he take them for… complete fools? He was brushing by them. Was it the lack of respect or the hollow sound of justification that bothered them more? They had heard the confessions of thousands of people. Their lives were given to forgiving. They daily counseled people who had done exactly what this man was confessing to. Teenagers had a greater sense of remorse for a first-time fling than this man did for ruining a church and many lives.

"That's it?" Heads turned to see Terry. His face was as red as the lights on the police car had been. "That is all you are going to say? You have nothing to say about the years of arrogance? You have nothing to say about the times you have acted or spoken against us? You have nothing to say about prayer for Lisa or the kids? You ask us to forgive, but you show no real remorse for your actions. Do you realize how this affects anyone beyond your little perch?"

"Easy Terry," Asa enjoined. "This is new territory for Robert, let's give him a chance."

"Terry, let's not let our personal differences cloud this issue." Robert Greene stood to address his former associate. "I have spent the morning with my family, Asa and George, and the elders of our church. They have formed a "Life Circle" around me and I invite you to be a part of that. There are serious issues in my life which I am not able to see clearly. This group has offered to work with me to help me find restoration. I am very embarrassed at being in this position. I do not know what the long-term consequences will be, but I ask you all to pray for me and to help me make it back."

Heads turned to Terry. He had become the default "offended brother." He did not look like he was going to release Robert until he was fully satisfied. Terry had spent six years as the Youth and then Associate Pastor under Robert. It had been his conviction that Robert was having an affair with his secretary. This conviction had caused him to seek out Asa's counsel, quietly leave the church, and begin a small group. For five years he had never said to another soul what he felt he had discerned. Now, with Robert Greene vulnerable, he had several very direct questions he would love to ask.

A strange calm filled him as he looked across the room to his one-time hero. It was a cup of kindness filled by hours of prayer for this man. He could destroy Robert Greene, the man, the myth, the legend. How his flesh had cried out for this moment. During the months that he had worked at the local food store because he had no severance package, he had thought how wonderful this moment would be. He had known that it would one day come.

The small group had become a "Life Circle" for him and Jean. Only three of the ten couples had been in the former church and they never talked about their time with or relationship toward Robert. Leading from within, Terry had helped them to develop vision and resources. They had decided that broken homes were to be their target and now long after they had begun strategic prayer for divorcees and their children, they had grown to over 150 people. They were able to tithe and free Terry and Jean to a ministry of reconciliation. In fact two of the men in this meeting were from restored marriages. Terry had learned to forgive. He had refused the cold stones of the pyramid for the warmth of the "Life Circle."

All this flashed through his mind as the calm swept over him. Would he throw that all away for one moment's glory? How he wanted to drag this man down. The calm cleared his thinking and, in the instant it afforded, he made his decision.

"Robert," the calm controlled his voice. "You know that the word 'Confession' means to 'Speak the same as' and you know that 'If we confess our sins He is faithful and just to forgive us.' The only sin that can be forgiven is that which is clearly stated for exactly what it is. Do you feel that your confession today fulfills those criteria for forgiveness? If I came to you as you have to us, would you feel that I had confessed in the sense of the word that could release forgiveness?"

"What was this?" Robert Greene was offended at being schooled by someone he had raised up in the ministry. Those were his words. He had taught Terry those very words and that approach in digging out detail so the person could see the hideous nature of what he had done and the depravity at its root. Why wasn't Asa asking the questions? How could he defer to this "Youth Guy?" This was more than he had bargained for. He began to plan his exit. Terry knew too much. If Terry kept asking questions, this was going to get very nasty.

Robert had always wondered just what Terry did know. The young man's silence was a great frustration to him. He could get anyone to open up. Robert was a master at getting people to tell him information he needed to rule the pyramid. When Terry had gone silent, Robert had lost his controlling edge. Now, he actually was feeling fear in his heart. What does this kid know that he has not disclosed?

The group waited for his response. Terry had been an example to them all. Each of these Pastors knew one or more families in their church who had come from Robert's. As they had come in, they had been very shy about the leadership format. Several had shared hurts and misgivings during their pre-membership meetings. Terry had deftly avoided all of their inquiries suggesting they go straight to the source and not him. They knew he had spent personal time with Asa and George, but respected his privacy as he had theirs. Asa was a rock. He gave no information. George responded to questions with a stare and his patented, "We're not going there."

Now they sat quietly and watched the color changes in the two men's faces. Terry had gone from crimson to sunrise pink. Robert was pale moving toward blue. He did not look well. Lips pursed with tension, eyes fixed on the younger man's face with serpentine focus, everything about Robert Greene said, "We are about to enter a life-and-death struggle." He was ready to attack.

"You would judge even my confession? Who are you? What have you done that qualifies you to judge me? If it weren't for me you would not even be in ministry. Half your church are people who could not handle the standards at our place. You are going to judge me? That will be the day." His outburst crackled through the room.

"Then, this is the day." George was about to stand. Their little drama had played for months in his prayer life. The Lord had led him through this scenario many mornings. He had not been surprised at Asa's call. In fact, he had been expecting it. It was the magistrate's call that had startled him. They were not close. In fact, he felt that the man gave favors to men who did not care financially for their estranged families.

When, at midnight, the call had come, he thought of going

down to the court, but the magistrate had suggested meeting Robert in the morning. The men had spoken of the possibility of the charges going away, but the magistrate said that several prominent men had been caught in the net and he was not in any position to help Robert. He felt that he could remove the felony conviction if Robert would serve some sort of probation period as determined by the church leadership. George had agreed to keep the man posted on developments on a confidential basis.

Robert's face went white as he looked at George. Even the great Robert Greene did not want to get into a war of words with this man. George's even gaze quenched all the fire in Robert's defense. "This is that day," George repeated. "Today you answer the questions of those who the community calls your peers. Whether you respect these men or not is of little consequence to them. But, it is the fundamental issue which brings you to this day in your life. Your lack of respect for the Body of Christ and her Head is the root of your sin. Today, you answer their questions."

His calm frankness and strength of conviction sent a message to Robert, "There is no game in this room."

Robert looked at the faces set before him. They were neither friendly nor compassionate. They were not angry, some seemed rather uncomfortable, but all of them had that professional calm of the ministry. They were one voice and heart. He was not going to divide their focus. They wanted the truth. Not that they were going to get any joy out of hearing the sordid details. They just wanted him to be honest with them and, more importantly, with himself.

Measuring the distance to the door, he stood and faced Terry. He was not going to do this. He was not going to humble himself before this or any other man.

"Pastor, don't do it," Terry appealed from his heart. "Don't run from it. Face the truth. Let these men help you."

He took one step back to clear the young man and, with head held high, walked out the door. Nodding to the busboy and narrowly missing the tray of fresh water glasses, he walked steadily down the corridor to freedom. He never looked back and they did not call to him. In one motion he swept open the door, and walked across the warm pavement to his car. This was it. He was not going to be in that bunch of losers again.

They reminded him of the glassy stares of the AA's who used his building. "Hello, my name is Robert and I am a whoremonger."

"Hi Robert." "Hi Robert." The reply became a chant in his mind as the motor roared to life. Pushing the accelerator to the floor, he narrowly missed the planters at the motel entrance.

Swerving into the traffic lane, with tires squealing, he launched himself in a straight run to the only haven he had known for months. Pulling the tie from around his neck as he drove, he narrowly missed an oncoming car. That would be a way out, he could choose a bridge abutment and just drive headlong into it.

The traffic light had just turned red, but it went unnoticed as he sped on his way to sin. Those men all wanted to do this. They were stuffed shirts. If he could not walk the walk, then there was no way they could. What a crew of losers they are. They couldn't possibly be the success he had been. They wanted glory, they wanted women, and they wanted all the money they could get. He deserved the finer things of life. What is he a monk?

"Hello, my name is Robert and I deserve the good life." "Hello Robert, Hello Robert, Hello Robert.

"Life Circle" my foot. I am not going to throw away 30 years of sacrifice and leadership to become a part of a bunch of men who could not cross the street without help. That is ridiculous. Not one of them can tie my shoes. They are losers and that Terry is the worst. I know he would love to be in this car with me headed to the babes and having a ball. That kid was nobody. I took him in when no one would have him. Now he is my judge, I don't think so.

"This is the day I declare my freedom from the whole mess. I am going to get a real woman; I am going to settle some money for my family; that's all they want anyway. I haven't fled the bed; I have been pushed out by three little gremlins.

"I didn't turn from Lisa; she left me for those kids. "Too tired," forget about it, "I'm going to find a woman who is never tired and who lives for me."

The car sped down the off ramp into the city center. Just blocks to go now and he would be in the arms of someone who understood his needs. "A pro," he thought. "A pro like me. Not a loser like those people."

He heard the ringing, but he couldn't find the phone. "Why hadn't he switched it off? Probably that Terry, begging him to come back to their little group. Wouldn't that be the limit? They could all announce that the fallen had come into their fold and the people would rejoice. Well, it wasn't going to happen and, as soon as he found that phone, he would tell him off for the last time." This was it. His anger was full throttle as was the car.

His hand landed on the phone and he snatched it from the cradle. "Why was the handphone in a cradle?" he thought as he fumbled with it.

"Robert?" It was Lisa. His office stared back at him. "Robert?" Where was the car? "Robert, are you there?" Eyes blinking rapidly he started from the chair. He wasn't driving. "Robert, are you alright?"

"Yes, yes, Lisa is that you?"

"Robert, sorry to interrupt, are you with someone?"

"Not just now." The room felt full of people: Asa, George, the

Elders, Terry, and wow what had happened to him?

"Honey, I just wanted to know if you will be home early for dinner?"

"Lisa, I will be home in about 20 minutes."

"OK, I love you."

"Lisa, I love you too. See you soon."

He dropped back into the chair. What had happened? Had it happened? Still numb from the emotions, adrenaline drain in his joints, he sat and stared at the desk. "Had it happened?" He did not know. It was so real. He must have fallen asleep. What had he been reading? There it was in front of him. The letter had been in the morning mail. Asa was inviting him to attend a Relational Group of local Pastors who had found strength in fellowship. They were exploring a new concept from a church in Asia called "Life Circles." His name had come up in the group as someone who the others thought could really flourish in such a vision for the city.

They were not asking him to lead. They were just inviting him to sit in with them and see how it went. As a courtesy the names of the others were included. Terry Barnes' name was third on the list.

His thoughts were interrupted by a knock at the door.

Mary entered at his greeting. "Will there be anything else today, Pastor?" she asked.

Looking back to the desktop he answered, "No thank you."

"I'll go then, if you are sure."

He lifted his eyes to meet hers. "Thank you Mary, have a good night."

As the door closed behind her, he buried his face in his hands and wept. It was her face. The woman in the car had been Mary.

Pastor Robert Greene picked up the phone and dialed the

number at the top of Asa's letter. It rang a few times and then he heard the older man's pleasant greeting.

"Asa, this is Bob Greene. Thank you very much for the invitation. I look forward to being with you at the meeting. Would there be any chance that you, and I, and perhaps George could have lunch the day before? No, no problem with Terry being at the meeting, I just would like to share with the two of you a very interesting thing that has happened to me. Tuesday at 11:30 will be fine and thank you very much for the invitation. Yes, God bless you also."

*Chapter 20*

# Sustaining Through Forgiveness

"OH," you say. "You let the man off with a warning from God. It was a dream. Did he do it?"

Certainly, as much as he did it in his heart, he did it. The signs are there for months before the actual event occurs. Robert Greene is no different than you and me. We each have our area of vulnerability. As the pyramid of pressure inverted on Pastor Greene, he began to hunger for companionship. He was a lonely man having no peer in his own mind. Because of the situation with his daughter's illness he lost what he had once found in Lisa. What steps must he take to be sure that his dream does not become a reality?

Through his dream he knows several issues with which he must immediately deal if his ministry is going to be sustained. **Sustain** is the **"S"** in **BRIDGES**.

## Establishing Fellowship

First, he must establish fellowship in the "Life Circle," which has been opened to him. This is going to require him to allow Asa and George to take the forest of feelings out of his eye. Robert has the paranoia shared by so many at the top of the pyramid. He feels that the ministers in town are out to get him. They want to bring him down. He would probably be surprised to find that they think of him often, but have no ill will toward him.

## The Grace to Forgive

Second, he is going to have to find grace to forgive on three levels. He must first forgive God for his daughter's condition. His disciplines are eroded because he has a personal belief challenge. Each time the agony of her breathing fills the house, a dagger cuts into his heart. How can God allow this to happen? Caught in his own pain, Robert does not want to pray or worship. He can go through the motions professionally, but he does not any longer personally believe.

Asa and George will be able to help him to find peace in God. When bad things happen to good people, our faith is attacked. They have ministered to hundreds of people and can minister also to him. They must restore the truth of the sovereignty of God in his life.

His second area of forgiveness is himself. All the emotional self-justification in his dream indicates self-condemnation. Anyone who is condemning himself has the Devil for a partner. Satan will accuse until the real and imagined are no longer separable. Robert's dream was so real that he no longer knew where he was. He did not know if it had happened. When the phone awakened him, he could feel the adrenaline in his body. His anger was becoming a rage and he was, in his dream, headed to a destructive act. This is really dangerous ground. His self-accusation has blended with his paranoia and he is really starting to lose touch.

Should he seek professional help? Perhaps a visit to a good listener will help. Robert has come very close to smashing the Rock. Help must come from someone who understands the way in which the pressures have combined to fuel his fires of passion. However, he must forgive himself for the pyramid perch. If he sees someone who is high on a pyramid in his own practice, it will avail nothing. The result will be the same as for Pastor Jon, "I have sought counsel from trusted friends and they affirm my course of action." The "Life Circle" is his best option.

The third area of forgiveness which the "Life Circle" must help Robert address is his conflict with Terry. This was most vivid in the dream. In reassessing it, Robert will see that Terry was actually a very compassionate friend who, unable to climb that pyramid peak backed away to pray. Terry had the opportunity to destroy and did not. Robert will have to sit down with the former protege and work out their differences. From the dream we see that Robert will have to be painfully honest in forgiving and receiving forgiveness from this young man.

Establishing the bridge of forgiveness with Terry will bring him to the local church. Here he faces the task of disassembling the current structure and forming "Life Circles." He should contact Pastor Naomi Dowdy and schedule a time for him and his board and staff to attend or host a Local Church Breakthrough Seminar with Pastor Dowdy. She has crossed this bridge and can show others the way.

He must draw together his board and love them for who they are, not use them for who they can make him to be. From his dream he knows that he has misjudged them. They are not disloyal. They can be trusted. They would stand in the gap for him with the community. However, he also knows that they are about at their breaking point and have said that trust will require one year of selflessness to be re-established. He has a lot of bridge building to do there. The Local Church Breakthrough Seminar will help, but it will be the "Life Circle" dynamic that will sustain the church.

## Facing Up to the Challenge

He has another issue at the church. What to do about Mary, his secretary? Are his desires for her stemming from Lisa's preoccupation with Megan? Evidently not, because Terry saw it before Megan was born. Could he be in love with his secretary? She supports a young family from her salary at the church. He could neither make her the culprit, nor the victim. How was he going to handle this? Again, he thought of the "Life Circle." Perhaps George or Asa knew of a young Pastor who could use the help of the most capable church secretary in town. Was that fair? Did it address the issue of his desires for her? More questions for the "Life Circle."

Terry and Lisa were right. He had to say things exactly as they were. He could not push the issue to her and ship her off to another church. Even if she went on, he would have to be honest with himself and God about his sinful desires. Who else does he tell?

There would be no gain in telling the church. He will have to weave a new depth in his teaching. There will be many times when he will feel the desire to make a confession to the church, but this is a private matter. Public confession would make him feel better, but what would it do to the others in the Circle? He would be using the body for his benefit instead of leading from within for their benefit.

The biggest question was, "What about Lisa?" In the dream, she knew. Because of her love for him, and her prayer discipline, she chose to stand with him, but all the while, she knew. How was he going to face a wife who had shown such love to him, a wife who had drawn from years of trust to keep her bridge from breaking? How he longed to go back to those days when it was just the two of them. He would take her away to that B&B she loved so much and would walk the streets of that old seaside village and together they would face this challenge as they had all the others. That is, all but one.

He had to forgive Megan. He had to abandon his resentment of her control of their lives and build a solid bridge of relationship with her. Actually, he was afraid of her. He was afraid that she would die in his arms. He was afraid he would hurt her as he pounded her back. He was afraid that her illness would cause him to lose position in the sight of the people. He was afraid he would lose his wife's affection to this child. Everything about Megan fed his fears. He could not love his own daughter.

"Wow," you say. "That man is in a real mess." You would be amazed how many leaders in all walks of life feel as he does. Doctors become addicts and their wives alcoholics because of these pressures. Attorneys go to prison where wardens cannot rest because the pyramids are crashing down on them. Students lash out in violence against their friends, five-year-olds take guns to school and Presidents are promiscuous with subordinates. Perhaps Dr. Graham was right, we do need to consider the pressures.

It is not too late for Robert Greene. He has been very fortunate. God has spoken to him in a dream and he has an opportunity to respond. He must go from this experience and build a strong relational base. From within that base will flow spiritual, human and material resource to enable that group to make an impact on their community and the world. He will have to lead them from within through an emphasis on prayer and he must return to his disciplines of worship, the Word, and work.

Robert is about to learn much about evaluation. He will find that it is better to confess sin and be forgiven than to try to run and hide. His mettle will be tested. From his dream he knows that he will want to run, but the fact of Mary's face being that of the woman is terrible enough to keep him moving in the right circles. **His way back will be a series of smart goals and his accountability will be to the "Life Circle."**

## Going On With God

You can meet Robert. Just go to any church in the world and look into the eyes of the Pastor. If brokenness looks back, you have met a post-dream Robert Greene. But, if arrogance and uneasiness respond to your greeting, give him this copy of BRIDGES and begin to pray.

Look through the church for Bob and Lesley. They will be there somewhere, taking care of the building and encouraging the new believers. If you get the chance to spend some time with their small group, you will be able to meet the people for whom this book was written. They are building a relational base, they will have a vision and be able to resource it from within the group. They are praying without ceasing and are sharing the disciplines of the faith with joy. Worship with them will be really great.

They will be talking about meeting the needs of people at home and around the world and you will often hear the word "change" in their conversations. They are going on with God to the harvest fields. They will have answered the questions, when, where, who, how, and how much, and will invite you to be a part of a God-given vision.

Above all else, they have seen enough to be totally committed to the sustaining of the work of the Lord in their community. This will be very important to them. In fact, they will be determined to serve the Lord regardless of what happens with their Pastor.

# About the Author

Mark Geppert was selling insurance to a Baptist pastor in 1970 when the Holy Spirit touched his life in an irrevocable way. Under careful training, he began walking out the call of God in his life. Leaving business in 1973 he received Bible and ministry training in the Assemblies of God at Western Pennsylvania Bible Institute.

From there Mark and his wife Ellie went to Guatemala where for three years he directed the Guatemala Assistance Project. Their first son, Samuel, was born in Guatemala.

In 1979 God spoke to Mark to return to Pittsburgh, Pennsylvania, and serve the Church in the establishment of the Dayspring Bible Training Center which opened its doors in 1980. A second son, Matthew, was born in Pittsburgh.

Mark served as Co-Pastor, Missions Director, and Bible Training Center Director until 1988 when he accepted a call to serve as Vice-President and Director of Operation Unreached with the Association of International Missions Services in Virginia Beach, Virginia.

In 1991 Mark and Ellie established the South East Asia Prayer Center. Committed to "Creating New and Networking Existing Prayer Cells in South East Asia," this foundation has brought forth fruit in many nations. Much of their success has been accomplished through Mark's Prayer Walking Seminar which has been taught to more than 50,000 in Asia, both through live conferences as well as audio and video cassettes.

Now 50, Mark has lived the principles of prayer. His life is a testimony to the faithfulness of Jesus to those who walk with Him. The focus on Asia continues with works in many nations.

To contact Mark Geppert,
write to:

South East Asia Prayer Center
Post Office Box 127
Oakmont, Pennsylvania 15139

or email:

seapc@hotmail.com

Please include your prayer requests
and comments when you write.